Praise for *Peculiar*

"This is a mental OS upgrade—no reboot required. Smart, insightful, scrappy, and startlingly practical; you'll start quoting it before you finish it."

— Raja Rajamannar, Chief Marketing & Communications Officer and Founding President, Healthcare Business, Mastercard

"A fascinating book. It delivers on many fronts. A window into the culture of Amazon—which Schlaff describes as his "favorite frenemy," a treasure trove of fascinating anecdotes, both wondrous and critical, and tons of practical actionable advice for anyone."

— Heski Bar Isaac, University of Toronto Distinguished Professor of Economics and Finance, Fellow of the Royal Economic Society

"Before I joined Amazon, Robert gave me invaluable advice on how to navigate its culture and leadership principles. His insights shaped how I approached my role, and I've drawn on them ever since. *Peculiar* distills that same wisdom into a book—clear, practical, and surprisingly human. It's the guide for anyone who wants to be more Amazonian."

— Rich Rabinovich, Former Software Development Manager, Amazon.com Chaos Engineering

"Mr. Schlaff knows of what he speaks: the forceful clarity of customer-obsession, the BS it tends to attract from copycats who have not in fact drunk the Kool-Aid, and the best ways to understand Amazon's vision and discipline. This book is a must to understand the world we live in now."

— Peter Temes, Founder and President, Institute for Innovation in Large Organizations

Rob Schlaff's *Peculiar* pulls back the curtain on Amazon's one-of-a-kind culture. This humorous, sharp and engaging book breaks down the principles that fuel the company's success and shows you how to put them to work in your own career. Whether you're just starting out or a seasoned exec, it's a must read!

— Seth Goodman, Chief Revenue
Officer, Boost Payment Solutions

"Whether it's sports or business, culture often determines whether you achieve your goals. *Peculiar* offers a firsthand look inside the culture of a mega-company, revealing the systems and mindset that drive success. More importantly, it uncovers how these cultural strategies can be adapted to new environments—and highlights the risks of misapplying them."

— Tera Hofmann, Former Professional
Hockey Goaltender

"As someone who's written about magical cats and alien overlords with Excel spreadsheets, I thought I'd seen it all. But Robert Schlaff manages to make Amazon's leadership principles feel like both a mythic quest and a practical survival guide. *Peculiar* belongs on your shelf—right next to your strategy decks and your fantasy novels."

— Sameer Chopra, author of
Captain Kitty Returns and *The Work Ahead*

Peculiar

Published by Daedalus Workshop 2025
Copyright © 2025 by Robert Schlaff
ISBN 979-8-9940818-1-5

The contents of this book represent the opinions of the author and in no way reflect the positions or opinions of Amazon.

A note on the cover: The punctuation mark in the cloud is an *interrobang*—a combination of an exclamation point and a question mark. It was proposed in 1962 by advertising executive Martin K. Speckter in *TYPEtalks* magazine, where he also crowdsourced names for the new mark. The interrobang became so popular in the 1960s that some Remington typewriters even came with their own interrobang key.

Peculiar

Lessons from Amazon's
Leadership Principles

Robert Schlaff

TABLE OF CONTENTS

A Peculiar Prologue

It was my first interview at Amazon. I was on the phone in an empty 15th-floor conference room, looking down at the Brooklyn Bridge. Things were going well.

"One final question," the interviewer said. "Why do you want to work at our peculiar company?"

I paused. "Wait, what?" I asked him to repeat the question. I had never heard Amazon described that way. Powerful? Yes. World-renowned? Of course. But peculiar? Never.

"What exactly do you mean by 'peculiar'?" I said.

He told me that Amazon is proud of being peculiar. In fact, peculiarity is one of its strengths. It's even in the introduction to the Leadership Principles, the core of Amazon's culture.

Amazon's peculiar strengths often get lost in translation. Here's one example. At another company, I received the following email.

Hi Coworkers!

It's your friendly neighborhood culture hackers. We're here to help you think outside the box and make our company like the world's best tech companies. We are here to share some killer ideas you can use to drive innovation. Here's one example for you!

At Amazon, in order to get fewer people saying "No" and more people saying "Yes," Jeff instituted the "Institutional Yes." Whenever a
new idea is suggested, the default answer must be "yes"—biasing
the company toward pursuing new ideas. If a manager wants to
say "no" to a new idea, they are required to defend their position
on Amazon's Intranet.

Try it out and tell us how it goes!

The Culture Hackers

I got this email while working at a large financial services company.
It came from a well-meaning group calling themselves the Culture
Hackers—a team I'd never heard from before and wouldn't hear from
again. Like so many others, they were trying to borrow what they
could from the big tech playbook.

I looked at this hack and thought, "Oh, that's interesting. I didn't
know Amazon did that." I wondered if we should implement that here.
Then I stopped and realized, "Wait! What?! I worked at Amazon. This
never happened."

Not only was this process made up, but it never could have existed
at Amazon. Why? Because that's not the way Amazon works. Amazon's
culture is defined by its 16 Leadership Principles. These Leadership
Principles form the backbone of this book. Asking a person to write a
memo about why they didn't move forward with an idea goes against
many of these principles. It violates the principles of **LP #10: Frugality**
(everyone at Amazon should think of time as a scarce commodity) and
LP #13: Have Backbone; Disagree and Commit (discussions about the
merits of an idea should be heated and interactive). Most significantly,
it's not about **LP #1: Customer Obsession**, the most important of all
Amazon Leadership Principles.

I sent an email to the Culture Hackers team. I told them that
while Amazon does have an "Institutional Yes," the process that they
described was fictional and created from whole cloth by a writer at

Entrepreneur magazine.[1] I didn't hear anything back. The team was probably embarrassed and wanted to brush it under the rug. The team members desperately wanted our company to be like Amazon. But here's one key difference between Amazon and most other companies: Amazonians (what Amazon employees call themselves) wouldn't have brushed it under the rug. They're not afraid to make mistakes, fix them, learn from them, and move on.

Jeff Bezos, Amazon's founder, sees Amazon's willingness to fail and learn as one of its key competitive advantages. A few months into my tenure at Amazon, I found some errors in Amazon's Leadership Principles. As Amazon's bible, the Leadership Principles are inculcated into every employee and even posted on the wall. As I discuss later in **LP #7: Insist on the Highest Standards**, I found typos in the Leadership Principles that had gone unseen for over a decade. When I emailed people internally, they said, "This is the funniest thing I've seen all day. Let's figure out how to fix it." Once I found the people responsible for maintaining the Leadership Principles, they were fixed in a matter of days.

I wrote this book to explain what Amazon really is, not what people imagine it to be. People assume Amazon makes fewer mistakes than other companies, but it actually fails more often. It tries more things, makes more mistakes, and learns faster. From the outside, it looks like a place where new businesses materialize out of thin air, but the reality is far less glamorous. Amazon isn't built on magic; it's built on an unyielding commitment to Customer Obsession and execution.

At the heart of it all is a culture that's deeply ingrained within the company but rarely understood by those looking in. That culture is shaped by the Leadership Principles. These principles aren't suggestions—they're how decisions get made, how teams operate, and how failure is treated as a necessary step toward progress. This book isn't just about understanding Amazon; it's about taking those

---

[1] Salim Ismail, "3 Ways Companies Can Encourage Smart Risk Taking," *Entrepreneur*, October 16, 2014, https://www.entrepreneur.com/growing-a-business/3-ways-companies-can-encourage-smart-risk-taking/238543.

same principles and applying them in your own career, your own choices, and your own way of thinking.

The Best Job in Banking

I'd always wanted to work at Amazon. It felt like the corporate big leagues—a place that was always a few steps ahead. When I joined, it felt like I'd arrived. I'd reached my goal: working at one of the world's most admired tech companies.

In June of 2019, I traveled to Japan for the first time. I was speaking at the AWS[2] Summit in Tokyo. Over 10,000 Japanese coders streamed into the Makuhari Messe Conference Center in suburban Tokyo to hear the latest AWS news.

I had the best job in banking. As the global Head of Banking for Amazon Web Services (AWS), I was Amazon's point person on cloud banking. Nearly every financial institution, from startups to global banks, used AWS in some way or was planning to. My job was to show them the future—to help them see what was possible and say, "If you use AWS, this could be you!"

Navigating through the flood of attendees, I felt I was in the game *Super Mario Bros.*, and I'd just been dropped into a water level—momentarily disoriented, but realizing that despite the unfamiliar setting, I still knew how to play. Everything was familiar yet slightly different. Searching for my area, I spotted a Japanese host wearing a T-shirt that said, "ASK ME! I'm with the AWS Summit!" When I asked for directions to the financial services booth, he responded with all the English he knew: "AWS! Yes. Yes. AWS!"

I was eager to immerse myself in Japanese culture. Even familiar items like cheesecake took on a magical new character—fluffier and sweeter than its American or Italian counterparts. Convenience stores

---

[2] Amazon Web Services, or "the cloud," is the ability to rent IT infrastructure. Instead of buying, owning, and maintaining physical data centers and servers, companies can access all the computing power they need on an as-needed basis.

like 7-Eleven were gourmet havens offering delicacies such as beef teriyaki jerky and dried squid.

While my hotel room had one tiny bed, the hotel had five bathhouses. These were traditional Japanese bathhouses. I had to try them. The signs said that visible tattoos and bathing suits were strictly forbidden. Each station offered unique hot and cold water experiences—one with massaging rollers, and another where the overflow cascaded majestically into a series of others. It was a novel and exciting theme park for nude cold plunges. At the same time, I was terrified that one of my business colleagues would come in and sit next to me. Luckily, the bath was empty the whole time I was there. I was in a world of sensory overload, constantly wanting more. If the 7-Eleven was this impressive, the best thing in Tokyo must be mind-blowing. According to my host, the best thing in Tokyo was the Imperial Palace.

Living Like an Emperor

The Imperial Palace is the main residence of the Emperor of Japan. After crossing the moat that once protected it from ancient invaders, I stepped into a world of history far older and more powerful than I imagined. I walked through a grassy lawn where the Emperor housed his concubines and visited the base of the Tenshu tower, which burned down in 1657. The rulers of Tokyo were so powerful that they didn't think they needed to rebuild it, thinking they would never be attacked again.

This was supposed to be the best place to visit in Tokyo, yet I felt something was missing as I wandered through the Palace's East Gardens. While the gardens were beautiful, they didn't seem vastly different from those in Central Park, just a few blocks from my apartment in New York. It felt pretentious, even silly, but I wanted more from these trees and plants.

Then I had a thought: how could I have a better experience? The Emperor had done his part. In 1968, he opened the gardens to the public, offering people like me the chance to enjoy his garden—except on Mondays and Fridays, when it was reserved for the Emperor and the Imperial Family to stroll around.

I wondered what the Emperor did on those private days. Could I do that too? I decided to sit and meditate next to the Emperor's iris garden, one of the most beautiful in the world. The irises were transplanted from the garden of Meiji Jingu, a shrine dedicated to Emperor Meiji, the great-great-grandfather of the current Emperor.

As I sat for an hour, my perspective totally changed. Instead of constantly seeking something more, I was able to appreciate what I already had. A sublime sense of calm and happiness settled over me. Strange and wonderful things began to unfold. A couple sat next to me, one of them wearing a Yankees cap. I soon learned they were from Chile and were in Japan visiting a friend they met through an international friendship organization founded by Jimmy Carter. The Yankees cap was a gift from their friend who lived in New York.

That hour in the garden offered a clarity I hadn't expected. I'd spent years climbing—always reaching for the next rung, the next title, the next company that might finally feel like enough. And I could've kept going. There were still higher rungs to climb. But sitting there in front of the irises once reserved for an emperor, I found myself wondering if it was worth it. I realized it was better to be like the Emperor on his days off—to walk the garden, take in the stillness, and remember that the best things in life are already there, if I just take the time to appreciate them.

Going Forward

Working at Amazon changed how I see the world. I used to think of it as a mythical company—somehow able to win in nearly every industry it entered. But over time, I came to see that its real advantage isn't magic. It's the discipline of its Leadership Principles and an unshakable obsession with its customers.

This book is an introduction to that culture. The principles I learned at Amazon became a practical toolset—one I've carried with me ever since. They're not silver bullets. You can't apply them by formula and expect instant results.

But when approached with intention and adapted to your own world, they become something more: a way of making sense of complexity. They offer guidance through uncertainty and a way to clarify priorities.

Working at Amazon gave me an incredible toolkit for getting things done. But it also made me realize that not everything in life is a problem to be solved. That moment in the garden didn't teach me how to scale faster or write a better six-pager. It taught me that I'd been a dog chasing the proverbial mail truck for so long, I didn't realize there was any other option. And that slowing down was my own way of being peculiar—one that worked for me, even if it didn't match Amazon's definition.

Amazon Basics

On my first day at Amazon, I walked toward a futuristic, *Blade Runner*-esque landscape on 34th Street, next to the Hudson River. This city within a city, Hudson Yards, is a glass colossus that sprouted up from an old train yard. I was heading into 5 Manhattan West, a building on its eastern edge. Once a brutalist concrete structure that housed the headquarters of Channel 13, New York's PBS station, it had been resurfaced to match its sleek new neighbors. The rent became too high for PBS, and now it was home to Amazon's advertising business. As I stepped onto the seventh floor, I found an industrial space with 15-foot ceilings, passing by small cafés and pastry shops in an environment that felt like a mix of steampunk and Epcot. I walked past a large indoor "Central Park" area covered in AstroTurf and dotted with beanbag chairs, where Amazonians had gathered to watch a quarterly All-Hands meeting that Jeff was leading from Seattle. The setup felt organic, providing an informal way for small teams to connect.

At the time, Amazon had over 5,000 employees in New York, but everything was decentralized. There wasn't even an auditorium large enough to bring everyone together when Amazon announced that New York was selected, and then de-selected, as Amazon's second headquarters.

Walking into orientation, I met people from many different companies and industries, all starting jobs across Amazon—from Amazon Advertising to Amazon Web Services (AWS). Some were joining divisions I hadn't even heard of, like Comixology, Amazon's comic book subsidiary.

About half of the new hires were headed to AWS, and many had come from Microsoft Azure, AWS's biggest cloud rival. During orientation, you could sometimes spot them tapping at their screens in confusion. "Tap. Tap. Tap. I think my computer's broken," they'd say, before remembering they were no longer using Microsoft Surface laptops with touchscreens.

Jeff the Magician

Jeff plays a central role in this book—the Leadership Principles are essentially his personal philosophies, scaled across the entire company. Throughout this book, I refer to Jeff Bezos, Amazon's founder, simply as "Jeff." Not because I know him personally, but because that's what all Amazonians call him.

At each All-Hands meeting, Jeff highlighted some of his favorite posts from Amazon's internal website. Once, he shared a humorous post from an Amazonian named Jeff that said something like:

> I am the founder of the Amazon support group "Jeffs Who Are Not Jeff." We come together to support the "other Jeffs" at Amazon. We meet every Thursday at 8 p.m. between the groups "Fire Phone Owners Anonymous" and "Amazonians Named Alexa."

Jeff and Amazon have transformed the world so profoundly that it's hard to remember life before the company existed. During my first weeks, I watched an early Amazon advertisement that showed what the world was like before Amazon:

A British teenager walks into a record store.

"Hi there," he says to the middle-aged manager.

"Hi," the manager replies.

"I'm looking for a CD for my girlfriend," the customer says. "I bought her something from you two years ago, but I forgot the name. It sounded like a combination of the Pixies and the Sex Pistols. She really liked it. Do you have another of their albums?"

The record store manager stares at him, bewildered.

It's easy to forget how much effort used to go into simply finding the right book, album, or movie. A search used to be an adventure—relying on word-of-mouth, magazine reviews, and searching through the aisles. Today, we barely think about it. The right choice appears almost before we realize we need it. In the world of that ad, shopping on Amazon would feel like magic—like a genie fulfilling your every shopping demand. But in reality, it's the result of technological innovation, operational excellence, and relentless customer obsession. How can the rest of us create that kind of magic?

Tech Companies Aren't Magic

Today, every company wants to be a tech company. Banks, movie studios, and even fast-food chains—everyone is chasing the digital transformation dream. And to be fair, there's logic to it. Technology makes businesses more efficient, scalable, and responsive to customer needs. But beyond the practical benefits, "tech" has become an identity—one that signals innovation and attracts rewards. It's not enough to run a great business anymore—you have to convince the world you're reinventing it.

Take McDonald's as an example. At its heart, McDonald's is a hamburger company or, if you read its mission statement, a company that "makes delicious feel-good moments easy for everyone." But that hasn't stopped them from trying to frame themselves as a tech player. They acquired an AI company for drive-thru optimization, rolled out automated ordering kiosks, and launched a mobile app that tracks your every McCraving. But at the end of the day, no one is ordering a Big Mac because McDonald's has cutting-edge AI tech. They're ordering it because they are five hours into a road trip and the kids won't stop hitting each other until they get a Happy Meal.

Many businesses believe they can transform into tech companies simply by copying what they see on the surface. One of the first things they try to emulate is the casual dress code. Dress codes signal culture—whether it's IBM's white shirts and black ties or the argyle sweaters and gray pants of prep schools. Many assume Silicon Valley's relaxed style began with Steve Jobs and the hippie counterculture of the 1970s. But the real origins go back further—to the godfather of Silicon Valley, Robert Noyce.

Robert Noyce, co-founder of Intel, was born in Burlington, Iowa, and shaped by a Midwestern Congregationalist work ethic. When he started Intel—the first modern tech company—he brought those values with him.

Noyce believed that no one was better than anyone else. He was the first to have a casual dress code in the Valley. But it wasn't about fashion or rebellion; it was about fostering a culture of meritocracy where the best ideas won, rather than the ideas from those with the fanciest suits and biggest offices. As other tech companies emerged in Silicon Valley, they inherited this approach from Intel.[3]

Most companies trying to mimic Silicon Valley's culture miss this crucial detail. They adopt the look without adopting the philosophy. But true innovation isn't about wearing jeans instead of suits. It's about building a culture where ideas rise on their own merit, regardless of who pitches them.

---

[3] Tom Wolfe, "The Tinkerings of Robert Noyce: How the Sun Rose on the Silicon Valley," *Esquire*, December 1983, pp. 346–74.

Behind the Curtain

Like any great magic show, we're compelled to ask, "How do they do it?" That's what I spent my orientation—and the next three months of transition[4]—figuring out. We were immersed in Amazon's Leadership Principles and learned how to apply them to our daily work. I finally got to see how the pieces fit together.

From the outside, Amazon looks like a holding company—a collection of seemingly unrelated businesses, from a bookseller to a grocery store to a television production company. Then there's my corner of the company, Amazon Web Services (AWS), the world's largest cloud provider. But what truly binds all these businesses together isn't corporate ownership—it's something far more personal: Amazon's culture.

Becoming part of this culture—drinking it in, living by it—is what it means to be Amazonian. At the heart of being an Amazonian are 16 Leadership Principles. These principles aren't just empty slogans; it's the operating system that runs Amazon.

When I joined, there were 14 Leadership Principles, beginning with **LP #1: Customer Obsession** and ending in delivering results. Everything in between was about *how* you got there. Two more principles were added after I left. In the rest of this book, I'll unpack these principles and show you how to apply them.

Oddly enough, these principles are central to Amazon's culture, yet they aren't kept secret. Right after orientation, I attended my first client meeting—a full-day executive briefing covering a wide range of topics. One session, titled "Amazon's Culture of Innovation," offered customers a behind-the-scenes look at how the company operates.[5]

As I listened, something clicked. The content was almost identical to what I had been learning as a new employee. I couldn't help but

---

[4] At many tech companies, the first three months are about understanding the company and how things work before starting "real" work.

[5] Amazon's culture has changed very little since it started. You can watch an early video of Jeff from 1999, when he had hair and a giant guffaw of a laugh, and hear basically the same things. Or read his first letter to shareholders that's distributed annually with the new letter.

wonder, *Are we really just giving away our secrets?!* It felt like we were oversharing, opening the proverbial kimono a little too far. But looking back, sharing Amazon's secret sauce with its customers was the point.[6]

Here's the catch: knowing Amazon's Leadership Principles is only a small part of implementing them.

Assembling the Puzzle

Before Amazon, I worked at the consulting firm Booz Allen Hamilton and held a Top Secret Clearance. It was serious business—certain electronics, even cans of Coca-Cola, weren't allowed in classified areas of the building.[7]

I learned that classified information comes in different forms. Some things are secret because of their content—like how to build an atomic bomb. Other information is already out in the world, but the real secret comes from how the pieces fit together. This concept is known as the mosaic effect—assembling scattered pieces into a larger, more meaningful picture, like creating a mosaic from small tiles.[8]

Understanding Amazon is like that. You can gather fragments—the Leadership Principles, the culture, the history—but to truly grasp how it works, you have to put the pieces together in the right way.

That's why I'm here. I worked with these principles every day at Amazon, and I've learned they're not one-size-fits-all. Like a mosaic, their meaning depends on how you put the pieces together.

---

[6] You can see this presentation yourself by searching YouTube for "Amazon's Culture of Innovation."

[7] Top Secret information could only be handled in a special room called a SCIF (Special Compartmentalized Information Facility). No electronics or cell phones were allowed in the room, so all cell phones were left at the door. At the time, Coca-Cola was having a promotion where a "surprise" can of Coke would tell the lucky customer that they won a prize! This led to signs on SCIF doors that said something like, "No unopened cans of Coke may be brought into this room due to the possibility of electronic surveillance from these cans." For further details on SCIF protocols and security measures, refer to NBC News, "Coca-Cola promotion prompts security measures," last modified July 1, 2004, https://www.nbcnews.com/id/wbna5345132.

[8] Office of the Director of National Intelligence, "Office of the Director of National Intelligence Classification Guide (ODNI CG)," last modified April 24, 2015, p. 22, https://www.intel.gov/assets/documents/702-documents/declassified/DF-2015-00044%20(Doc1).pdf.

Tenets of This Book (Unless You Know Better Ones)

Jeff often says that Amazon is "stubborn on the vision, flexible on the details." The specifics may shift, but every project starts with a clear sense of direction. To put this into practice, teams define a set of key tenets—concise principles that shape priorities and guide decisions. These tenets act as a north star, keeping the big picture in focus even as plans evolve. By holding on to them, teams stay aligned with the long-term vision while leaving room to adjust, refine, and find the best way forward.

Here are some tenets I used when writing this book:

1. **Make the Leadership Principles Accessible.** Use stories—from Amazon and from my own life—to make the principles real and relatable.
2. **Be Interesting and Practical.** This isn't a book about theory. It's a hands-on, sometimes sideways look into Amazon's culture and how it actually works.
3. **Inspire Action.** Show how powerful the Leadership Principles can be—and offer practical ways for readers to apply them in their own lives.
4. **Make It Fun.** Remember that this isn't a textbook and I should never take any of this too seriously.

Now, about that "unless you know better ones" bit at the top of this section. It's included in most Amazon tenets as a kind of permission slip to challenge what's already there—an acknowledgment that no rule is sacred, no process immune to scrutiny. Good ideas can come from anywhere, regardless of rank or tenure.

You might think that at a trillion-dollar company, processes would be set in stone, with employees simply following established rules. But "unless you know better" exists to prevent that kind of complacency.

Without it, people might go through the motions, following outdated practices just because that's how things have always been done.

This kind of pattern—repeating forms without understanding their function—isn't just a corporate habit. It's something we all do. Richard Feynman called it the "cargo cult mentality," after islanders in the South Pacific who, during World War II, saw cargo planes arrive full of supplies. After the war, they tried to bring the planes back by recreating what they saw during the war: runways, signal fires, even bamboo headsets. Looking from the outside, it's clear it wouldn't work, but they couldn't know that. They were doing their best with what they knew—repeating the process that they saw before, hoping to get the same results.

Amazon strives to avoid this kind of thinking. Instead of blindly following tradition, employees are encouraged to question processes and improve them—they just may know better.

Stepping Inside

Now, let's get started and put you in the middle of the action. Most books about Amazon and other tech companies treat the reader like a spectator. The companies are amazing, almost mythical. It's like watching *Charlie and the Chocolate Factory*—a world filled with wonder and eccentric brilliance. In fact, the UK publication *The Register* even refers to another tech giant as "The Chocolate Factory" because it's as strange and awe-inspiring as Willy Wonka's world of candy.

And yes, there are many incredible things about Amazon, and I'll share those with you. But simply marveling at them won't tell you what it's actually like to be inside Amazon.

What if, instead of just watching the movie, you could get inside the mind of Augustus Gloop—the gluttonous boy who fell into Mr. Wonka's chocolate river—and experience firsthand how the adventure changed him? That's the perspective I want to offer in this book.

So now, let's step inside Jeff's peculiar company.

A Peculiar Company

As I mentioned in the Prologue, Amazon is proud to be peculiar—and that peculiarity is one of its greatest strengths. Jeff likes to say, "As we do new things, we accept that we may be misunderstood for long periods of time." It's not just branding. Peculiarity is a core part of Amazon's culture.

The company even has a peculiar unofficial mascot named Peccy—though I've rarely seen it used. Imagine an orange peanut M&M that got squished slightly, sprouted stubby limbs, and started beaming at you with wide, excited eyes.[9]

But it's not peculiarity for peculiarity's sake. Amazon focuses on making the best long-term decisions, even if they might feel uncomfortable or seem strange to others in the short term. This can make starting at Amazon a jarring experience. It certainly was for me.

I remember walking into my first Amazon Document Review meeting. I was in a boardroom with 20 people gathered around a large rectangular table, ready to learn about a potential new product.

---

[9] Michael Grothaus, "Amazon's Mysterious Mascot Peccy Is the Star of a Thriving Underground Merch Market," Fast Company, April 12, 2019, https://www.fastcompany.com/90329525/amazon-peccy.

I looked to the front of the room, expecting to see a presenter with a PowerPoint projected on the screen. Instead, everyone was staring at a paper in front of them, reading silently.

I grabbed a seat, wondering when the meeting would actually start. Then someone handed me my own copy of a six-page document and a pen to start marking it up.

This peculiar silent reading is a hallmark of Amazon's Document Review process. Instead of a presentation, the first 20 minutes are spent silently reading the document and making notes. Once everyone finishes, the group discusses the content and offers feedback. Finally, the author collects the comments and incorporates them into the next version of the document.

Jeff introduced this process in 2004 because he was frustrated with the PowerPoint presentations at his senior meetings (called S-Team meetings). He noticed that his senior leaders had stopped discussing and debating ideas. Instead, they were using their presentations to sell their ideas to each other with fancy graphics and compelling speeches, rather than coming up with the best possible solutions. So he banned PowerPoint and forced everyone to write up their ideas as short text documents called narratives.

Where did Jeff get this idea? Inside Amazon, I heard that he grew bored while listening to yet another PowerPoint presentation. He noticed that the speaker had a set of papers full of talking points that could be used in case any difficult questions came up. Jeff asked to see the papers and found them far more thoughtful and useful than the presentation. That became the heart of Amazon's writing culture. But that story is a myth.

The real story, as told by Colin Bryar, Jeff's Technical Assistant,[10] is that Jeff got the idea when reading Edward Tufte's **"PowerPoint Does**

---

[10] Jeff's Technical Assistant, also known as his shadow, is a key role at Amazon, often filled by a rising leader who works closely with Jeff to gain insights into the company's operations and strategic thinking. This position allows the assistant to understand Jeff's decision-making processes and the intricacies of Amazon's business, providing a unique opportunity for mentorship and leadership development within the company.

Rocket Science–and Better Techniques for Technical Reports."[11] Tufte argues that PowerPoint is not a decision-making tool but a sales tool. It shortens ideas into brief bullet points designed to sell an idea, often smoothing over complexity in favor of something that sounds great but lacks substance.

PowerPoint is excellent at making the presenter look good, regardless of the quality of the underlying idea. But inside a company, people should be discussing ideas, not selling them to each other. To illustrate PowerPoint's flaws as an internal communication tool, Tufte highlighted its role in poor decision-making at NASA. He pointed to a slide from engineers warning of risks before the Columbia shuttle disaster—so poorly structured that its most crucial message was lost.

Amazon's Document Review process is far better than PowerPoint for making decisions. Reading a six-page memo replaces the PowerPoint presentation at the beginning of the meeting. Documents are capped at six pages—just enough that, at three minutes per page, the first 20 minutes are spent reading. This replaces the 20 minutes typically spent presenting. The remaining two-thirds of the meeting is used for discussion and refinement, often generating over 100 comments. Afterward, the document serves as a permanent record of the idea and can be read by those who weren't at the meeting.

You might ask, "Why don't people send these documents as pre-reads before the meeting? Wouldn't that be more efficient? You could save so much time!" This is where Amazon's peculiarity really shines. It doesn't pretend that people will magically find 20 minutes in their already packed schedules. Since reading the document carefully is important, it sets aside time during the meeting for everyone to read the document together—even if that seems peculiar.

---

[11] Colin Bryar and Bill Carr, *Working Backwards: Insights, Stories, and Secrets from Inside Amazon* (New York: St. Martin's Press, Kindle Edition, 2021), pp. 85–86; Edward Tufte, "Power-Point Does Rocket Science—and Better Techniques for Technical Reports," accessed February 15, 2025, https://www.edwardtufte.com/notebook/powerpoint-does-rocket-science-and-better-techniques-for-technical-reports/.

The Amazon Way of Thinking

During annual reviews, Amazonians are asked, "What is your superpower?" As a company, Amazon's superpower is an X-ray vision to see through the noise and distraction and focus on what truly matters. It takes an honest look at itself and its work, refusing to chase trends or indulge in wishful thinking. Instead, it zeroes in on projects that will actually make a difference. Jeff put it this way:

> I very frequently get the question: "What's going to change in the next 10 years?" And that is a very interesting question; it's a very common one. I almost never get the question: "What's not going to change in the next 10 years?"
>
> And I submit to you that that second question is actually the more important of the two—because you can build a business strategy around the things that are stable over time... In our retail business, we know that customers want low prices, and I know that's going to be true 10 years from now. They want fast delivery; they want vast selection.
>
> It's impossible to imagine a future 10 years from now where a customer comes up and says, "Jeff, I love Amazon; I just wish the prices were a little higher," or "I love Amazon; I just wish you'd deliver a little more slowly." When you have something that you know is true, even over the long term, you can afford to put a lot of energy into it.[12]

Jeff is making the point that the core value proposition of Amazon is fairly simple. Everyone knows that customers want things cheaper and faster. But executing on that mission is difficult and requires the company to stay focused. Instead, companies often spend their time

---

[12] Jeff Bezos, "2012 re:Invent Day 2: Fireside Chat with Jeff Bezos & Werner Vogels," YouTube video, posted by "Amazon Web Services," November 30, 2012, https://www.youtube.com/watch?v=O4MtQGRIluA.

on easier but less important things. Back in the 1950s, naval historian C. Northcote Parkinson illustrated this with a now-famous story.

He described a committee tasked with reviewing two proposals: one for an expensive nuclear reactor, and another for a bike shed in a public park. Faced with the complex reactor, the committee stalls—it's too technical, too risky, and no one feels qualified to weigh in. So they push it aside. But the bike shed? Everyone has an opinion on that. They debate the color, the materials, and the locks in excruciating detail.

The story was so powerful that it earned its own term—bikeshedding. This pattern plays out in companies all the time. Rather than making the tough, high-impact decisions, teams get stuck obsessing over the easy, inconsequential ones. They'll even rationalize their inertia. I worked with one company that took years to figure out how to execute on a technology trend. After the trend had passed, one senior leader said, "See? It was good we took so long. We saved so much money by not doing anything!"

Getting Into Shape

Starting at Amazon is hard. It's not that anyone hazed me—the people were really nice. And it's not like I was working crazy hours—I've worked much longer hours in consulting. It was hard because I was out of shape. Specifically, I was out of "learning shape."

I wasn't used to this. At my previous jobs, the goal was to get my work done, follow the process, and meet my annual targets. At Amazon, I needed to do all of that and continually improve myself. It's like going to the gym.

When you go to the gym, it hurts. You're breaking down your muscles at each workout, and they come back stronger. When you're out of shape, the pain you feel in the first few days is the worst. Amazon was like that—but instead of a workout for my body, it was a workout for my mind.

At Amazon, "good enough" isn't good enough—at least not when it comes to serving the customer. In his 2016 shareholder letter, Jeff wrote: "Customers are always beautifully, wonderfully dissatisfied, even when they report being happy and business is great."

That's Amazon in a nutshell. It's: "I don't care if the customers are happy. We can still make them happier—or someone else will."

"How wonderful to be at a company that's always learning!" you might say. "I want to keep learning. I LOVE learning."

But when was the last time you truly learned something new? Maybe in college? You've probably forgotten how hard it is.

Learning is about getting B's and C's before you get an A. Most companies don't give B's and C's. You expect to get an A on every PowerPoint presentation you deliver.

The Best Place to Fail

A CEO once told me, "I want to build a new product that will generate $100 million a year in revenue within two years, require low investment, and be low risk." Those projects don't exist. And yet, many companies still try to innovate without accepting any risk of failure. That's where Amazon sets itself apart.

Jeff often talks about the trade-off between innovation and failure, as in his 2015 Letter to Shareholders: "One area where I think we are especially distinctive is failure. I believe we are the best place in the world to fail (we have plenty of practice!), and failure and invention are inseparable twins."

Jeff talks about failure as a key ingredient of innovation, but it's often misunderstood. Some companies treat failure as a goal that they can measure, assuming that if they fail enough, innovation will follow. I once had a boss who insisted, "If we're not failing a third of the time, we're not pushing hard enough."

But failure isn't something to strive for—learning is. The only way to learn something that hasn't been done before is to experiment. Trying something truly new is bound to fail sometimes. But if you're really doing something new, you don't know if it's going to work or not. The only way to learn is to try it.

It's like getting better at skiing. To be a great skier, you can't be afraid to fall. If you are afraid to fall, you'll never leave your comfort zone and you won't improve. You need to get comfortable with falling so you know how to do it without hurting yourself. But you also need to seek that good fall—the one where you go just beyond what you can handle, hit the snow, and get up knowing exactly what to adjust next time.

The Growth (Amazon) Mindset

Professor Carol Dweck has a good model for how tech companies like Amazon differ from other companies. She describes two mindsets: the growth mindset, which is similar to Amazon's approach, and the fixed mindset, which reflects more traditional, risk averse companies.

In a growth mindset, people believe their abilities can be developed through dedication and hard work—brains and talent are just the starting point. They are always learning, improving, and adapting to a changing world. Failure is seen as something to learn from, not as a judgment of ability or worth.

The simplest version of a growth mindset vs. a fixed mindset is from a cartoon by Sarah Andersen[13]:

> **Fixed Mindset:** How do you draw so well?
>
> **Growth Mindset:** Practice.
>
> **Fixed Mindset:** It must be an innate gift... A gift from God...
>
> **Growth Mindset:** It's practice.
>
> **Fixed Mindset:** I'll never understand how some people are so talented... A mystery...
>
> **Growth Mindset:** Practice.

[13] Sarah Andersen (@SarahCAndersen), X.com, December 20, 2017, https://twitter.com/Sarah-CAndersen/status/943504157960421377.

Dweck discovered the importance of failure early in her career when she studied how students handle tough problems. She watched children tackle a series of increasingly difficult puzzles. Some of them couldn't be solved at all. Expecting to see different coping strategies for how to deal with failure, she was surprised by the reactions of some students. One ten-year-old enthusiastically exclaimed, "I love a challenge!" while another, struggling with the puzzles, confidently said, "You know, I was hoping this would be informative!" These responses puzzled Dweck, who had always seen failure as something to bear or avoid, not something to welcome. This moment led her to explore the idea of a growth mindset, where failure is not a setback but a chance to learn and grow.

Jeff is a diehard believer in the growth mindset. He calls it "Day 1." At Amazon, Day 1 isn't about time; it's about mindset. It means always being in startup mode, always having more to learn and ways to improve. Each building at Amazon has a name associated with its history, and Jeff works in a building called Day 1. When he moved to a different building, he took the name with him.

This led to a lighthearted initiation ritual—one that, of course, I walked right into:

"Hi, Rob, how many days have you been here?" a coworker asked.

"It's day three for me," I answered.

"No, Rob. It's always Day 1 at Amazon."

Always trying to learn and improve can be terrifying. It's like flying on a trapeze without a net. Jeff and his ex-wife, MacKenzie, took this idea seriously when raising their kids. They wanted their children to create and learn as much as possible—even if this meant accepting some significant risks, like letting their kids use sharp knives at age four and certain power tools at seven or eight. When asked why, Jeff quotes MacKenzie saying, "I'd much rather have a kid with nine fingers than a resourceless kid."[14]

When Amazon has a goal, it has an almost bullheaded stubbornness to get things right, failing again and again until it does.

Learning from Failure: Amazon's Third-Party Selling

It's easy to see Amazon's successes as inevitable, but the journey to success is often riddled with failures. Take third-party selling, for example. Today, Amazon is the top platform for smaller merchants and manufacturers to sell their goods. Since 2014, Amazon has sold more products from other sellers than from its own inventory.

However, Amazon faced significant failures along the way. In 1999, it launched Amazon Auctions to compete with eBay, but it failed and was shut down in 2001. Then, Amazon introduced "zShops," mini-websites where merchants had their own sections. This setup was unpopular because it created a fragmented shopping experience— customers had to view a product on Amazon's site and then again on the merchant's site.

The problem was that customers came to Amazon to buy the thing they were searching for quickly and easily. When Amazon introduced Auctions and zShops, they didn't know what to make of it. The experience felt like a detour, not an improvement.

So Amazon made a risky bet. It gave third-party sellers equal footing—placing their products directly beside its own. It was a bold move, inviting competition into the heart of the business. But it worked. This unified approach caught on and set the stage for a thriving third-party business. If Amazon had given up after failing once, or even twice, it would be a very different company today.

[14] Jeff Bezos and Mark Bezos, "Amazon CEO Jeff Bezos and Brother Mark Give a Rare Interview About Growing Up and Secrets to Success," YouTube video, posted by "Summit," November 15, 2017, https://summit.co/videos/amazon-ceo-jeff-bezos-and-brother-mark-give-a-rare-interview-about-growing-up-and-secrets-to-success-3nBiJY03McIIQcgcoe2aUe.

Fixed (Typical Company) Mindset

Professor Dweck defines a fixed mindset as the belief that core qualities like intelligence, creativity, or talent are innate and unchangeable. When people hold this belief, they tend to see life as a constant test— each challenge a measure of their true worth.

At work, people with a fixed mindset categorize others into rigid roles. Computer coders have computer science degrees. Good businesspeople are successful and bad businesspeople are not. Hares run fast and tortoises will always lag behind.

People and companies with a fixed mindset are afraid to fail. Failure is seen as a mark against them, proof that they are less capable than before. These companies say, "Why should we promote someone who fails? We only want successful people here!"

Jeff calls the fixed mindset "Day 2." This is when a company stops learning. It is when rigid processes take over, preventing the company not only from growing effectively but also from responding to an ever-changing world.

Every company needs processes—even Amazon. Amazon hates the word "process." It avoids processes whenever it can, creating lightweight "mechanisms" when needed.

Processes exist to serve the business. They reduce rework, improve coordination, and instill best practices. But if you are not careful, the process itself can become more important than its intended outcome.

As Jeff noted in his 2016 letter, processes can sometimes get out of hand. Employees start focusing on following the process rather than achieving the actual goal. This can lead to bad outcomes, defended with, "Well, we followed the process."

That's why "best practices" can be dangerous when followed too closely. One company I worked with invested millions into a sleek, state-of-the-art innovation lab. They gathered some of their top innovators and proudly showcased the space on their website. Then they waited for the breakthroughs to roll in. Nothing happened. It turned out that building the lab wasn't the key to driving innovation. A beautiful, airy, column-less space might help people think—but it's a nice-to-have, not a need-to-have.

Final Thoughts

Amazon's peculiarity is key to its success. This mindset is not about being unusual for its own sake but about making deliberate decisions that prioritize long-term success over short-term comfort. To succeed, you must see the world as it is, even if others see it differently.

For example, if you want everyone to read a document, schedule time during the day for them to read it instead of expecting them to find time on their own. It might seem peculiar, but it works.

I've learned to apply Amazon's growth mindset at home by focusing on what actually works, not what looks good. That means letting go of how I think a "good parent" should act—or how "well-behaved kids" should behave—and instead adjusting, experimenting, and improving as we go.

One Saturday morning, we were getting ready to leave for a party at 9 a.m. I wanted to think of myself as a responsible parent with responsible children, so my wife and I told the kids, "We are leaving at 9 a.m. Make sure you are ready exactly at 9 a.m."

As we were getting dressed, I had a feeling things weren't going perfectly on the kids' side of the house. I turned to my wife and said, "I'm going to check on the kids. I know they're 9 and 12 and can do this themselves, but I just want to see."

Sure enough, I found them watching TV.

"What are you doing?" I asked.

"Watching TV. We are all ready to go," they said.

"Did you brush your teeth and put on your shoes?"

"No," they said.

"So you're not ready to go," I said.

It wasn't that they didn't want to be ready—they just needed guidance. At Amazon, good intentions aren't enough—you need mechanisms. That morning, telling the kids to "be ready at 9" felt reasonable. Clear. But it was the kind of directive that seniors give at work but don't know how to implement, like "We're moving to cloud," or "Integrate AI into everything you do."

Instead of getting angry at my kids for not being ready, I built a simple feedback loop: check in every 10 minutes, see how things are going, make small adjustments. It wasn't about catching them off guard or enforcing a rule. It was about working together to understand what "be ready to leave at 9" meant and how to achieve it.

It was annoying and didn't give me the nice "I have responsible children that listen to me" feeling. But it worked.

By building a culture that values a growth mindset and a willingness to embrace failure as part of the process, Amazon has become one of the world's largest and most powerful companies. This peculiarity, though sometimes challenging to adapt to, enables Amazon to bring together a wide range of businesses successfully.

Now that we have a better understanding of Amazon's culture, let's dive into the Leadership Principles.

Leadership Principles

Now, let's get into the heart of the book: the Leadership Principles. Amazon's culture is the bedrock of the company, and the LPs are its foundation. Amazonians often shorten it to LPs because they use the term so frequently. At many companies, culture is just a sales pitch—something designed to attract customers and employees but often relegated in favor of more pressing concerns like making as much money as possible. At Amazon, however, these principles define what it means to think and act like an Amazonian. They shape every interview, meeting, and performance review at the company.

New Amazonians tend to be skeptical when they first see them. Many ask, "These Leadership Principles are interesting, but do you actually use them after orientation? Don't they just get put in a drawer while you get to work?"

Of course, no one memorizes all 16 principles and methodically applies them to every action—but that's not the point. The LPs aren't about passing a quiz; they're more like a philosophy or even a religion. At Amazon, the principles become a blueprint for how you operate every day. They are reinforced constantly—until they become second nature. For veteran Amazonians, it's hard to understand why others wouldn't think this way.

It's like this story of two fish:

> Two fish are swimming along. They happen to meet an older fish swimming the other way, who nods at them and says "Morning, boys. How's the water?"
>
> And the two young fish swim on for a bit, and then eventually one of them looks over at the other and goes "What the hell is water?"[15]

The LPs are everywhere, woven into the fabric of the company. Walk down any Amazon office hallway and you'll see phrases like Dive Deep engraved on the walls. In New York, black-and-white photos of the city were displayed with LPs subtly hidden in store windows or on billboards. The principles are more than just words; they shape Amazon's DNA.

The LPs also serve another purpose: they scale Jeff.

When he started Amazon, it was easy. The company was small, it was just three people in a room, and Jeff was involved in everything from hiring candidates to shipping boxes. But as the company grew, it became impossible for him to be personally involved in every decision. So the question became, "How do you scale Jeff?"

He needed a system to spread his core values and high standards across the entire company. So he created the Leadership Principles. This wasn't about creating a culture—it was about codifying it. As Jeff wrote in his 2015 letter: "You can write down your corporate culture, but when you do so, you're discovering it, uncovering it—not creating it."

To develop the LPs, HR leaders interviewed employees across the company to identify the key qualities that made Amazon successful. Jeff and senior leaders then refined and adjusted the principles over several months, ensuring they reflected what Amazon already valued.

---

[15] David Foster Wallace, *This Is Water: Some Thoughts, Delivered on a Significant Occasion, about Living a Compassionate Life* (New York: Little, Brown and Company, 2009).

Be Careful When Using the LPs

The LPs are powerful tools, but they need to be used carefully. I like to think of them as sharp knives.

Sharp knives are incredibly effective, but they need to be treated with care. For people who don't know what they're doing, they can be quite dangerous.

I have a friend who was enamored with a knife she bought online. She said, "How could I not buy it? It was $10 for the sharpest knife in the world." She ended up throwing it away after a month—after constantly cutting her fingers with it. As I once heard, "They say they want sharp knives, but I don't want them running around with scissors."

I know of one company that tried to implement Amazon's Six-Page Narrative as a replacement for their quarterly review process. Here's all they knew about the Six-Page Narrative:

- Documents must be in Microsoft Word.
- They must be six pages long.
- No slides will be shown during the presentation.

So they shoved their existing process into Word, giving everyone a six-page Microsoft template to fill out and send around as a pre-read. But they missed what makes the Six-Page Narrative so effective.

A Six-Page Narrative isn't a template. It's an exercise in clear writing and thinking. And most importantly, it must be read silently during the meeting and then discussed—not sent out beforehand as a pre-read.

By implementing only parts of the Six-Page Narrative while stripping it from the broader document review process, the company made their meetings significantly worse. Teams struggled to cut and paste their PowerPoint templates into Word. No one read the documents beforehand, and during the meeting, there was neither time to read them nor a presentation to explain them.

The LPs aren't magic spells. They work because they force clear thinking, not because they follow a rigid format. Slapping a six-page

Word doc onto an old process doesn't make it Amazonian—it just makes it annoying.

Diving Into the Individual LPs

Before I worked there, I'd heard about Amazon's culture—scattered mentions in magazine profiles or blog posts, a clever tactic here, an unusual practice there. Many of these are repeated so often they have a name: Jeffisms. I was impressed by the clever and thoughtful way things were done. But when I arrived at Amazon, I saw how it all fit together.

The through-line was the Leadership Principles. They weren't just aspirational posters or HR-speak. Yes, there were a lot of posters—but they also showed up in real decisions, in how people talked, in what got rewarded.

Though I'd appreciated some bits and pieces of the culture before, I started to learn that the Leadership Principles weren't always what they seemed. Some sounded obvious but turned out to be surprisingly subtle. Others looked harsh on the page but were a lot more reasonable in practice. "Are Right, A Lot," for instance, isn't about winning arguments—it's about developing the kind of judgment that comes from being wrong and learning from it.

In the chapters ahead, I walk through each principle, tracing how I came to understand it—not just what it says, but what it actually does inside Amazon. I look at how each one plays out in the organization's day-to-day rhythms. Then I step back and ask a broader question: how does this work outside of Amazon?

Because the truth is, not all of them translate neatly. Amazon was built to move fast, scale endlessly, and make bold bets. Most environments—schools, banks, families—run on a different fuel. So I've tried to explore where the principles hold up and where they need a little adjustment.

Now, on to the LPs!

LP #1

CUSTOMER OBSESSION

Leaders start with the customer and work backwards.
They work vigorously to earn and keep customer trust.
Although leaders pay attention to competitors, they
obsess over customers.

I constantly remind our employees to be afraid, to wake up every morning terrified. Not of our competition, but of our customers. Our customers have made our business what it is; they are the ones with whom we have a relationship, and they are the ones to whom we owe a great obligation. And we consider them to be loyal to us—right up until the second that someone else offers them a better service.

—Jeff, 1998 Amazon Letter to Shareholders

During orientation, I heard the following story about Customer Obsession. Picture this: it's two days before Christmas. A manager is walking through an Amazon Fulfillment Center[16] and spots an unmailed Christmas present in the corner. It was ordered in plenty of time but somehow got overlooked.

---

[16] "Fulfillment center" is Amazon's term for a warehouse. Amazon likes to rename things to make them seem more friendly—like the way Disney calls customers "guests."

The manager springs into action, checking all possible shipping options to get the gift under the tree. From airmail to couriers, nothing can get it there in time. Then he realizes he's flying home from an airport just a half-hour from the recipient's house. He quickly grabs the package and delivers it personally on his way to the airport. He's never seen a happier customer.

What crap! I thought. *No one actually does this in real life.* But Jeff says Customer Obsession is the most important thing, so I figured I'd give it a shot.

That weekend, at a baseball game, I overheard that Chris, one of my son's teammates, wanted to dress up as an Amazon package for Halloween. This was my opportunity. I asked my orientation leader, expecting her to say, "Great, now get back to work." Instead, she was excited to help. She suggested a fulfillment center, but the closest one was an hour away. Then a friend mentioned there was a Prime Now[17] fulfillment center in the basement of our building.

I took the elevator down, expecting cubicles and foosball tables like every other floor in the building. Instead, I stepped into a maze of towering shelves packed with everything from cereal to toilet paper. A manager walked by. "Can I help you?"

"I need packing materials for a kid's Amazon box costume."

He handed me some bags and stickers. Problem solved. Thanks to a team effort, Chris got his costume—and I got my first real taste of Customer Obsession.

Working Backwards[18]

So how does Amazon take this customer-obsessed mindset and scale it? It starts with the customer.

---

[17] Prime Now was a one-hour delivery service of convenience store items. It's been discontinued as Amazon has increased its Prime shipping speeds and launched Amazon Fresh more widely.

[18] Amazon Web Services CTO Werner Vogels documented the process in 2006 and it's been virtually unchanged ever since. https://www.allthingsdistributed.com/2006/11/working_backwards.html

Most companies begin the product development process from the company's point of view. A product manager looks at market opportunities, projects annual sales, and calculates costs. They map out the high-level features and requirements, then hand it off to the tech team to build while sales and marketing work to convince customers how great it is. The actual value for the customer—the "what's in it for me?"—often doesn't get defined until the very end of the process.

Many years ago, a company I worked with set out to build an iPad banking app for CFOs to manage their corporate bank accounts. It wasn't just about showing global account balances—it even had a personalized news feed to keep them informed. The team spent months refining every detail, convinced it would be a hit.

But when it launched, CFOs didn't use it. As it turns out, they don't manage bank accounts. That's the job of the Head of Cash Operations, who reports to the Assistant Treasurer, who reports to the Treasurer, who reports to the CFO.

It never had a chance.

Amazon avoids this problem by starting with the customer. In the early days, teams would pitch Jeff big ideas, full of market data and revenue projections. But he wanted more. What would the experience be like? How would it make a customer's life better? If they couldn't answer that, the idea wasn't ready. It wasn't enough for a product to look good on paper—it had to matter to customers in the real world.

The company scaled Jeff's approach into a process called Working Backwards. This has a double meaning. First, Amazon works backward from the customer to build a product based on what the customer actually wants. It's also "backward" compared to how most other companies do it. Instead of starting with the business need and then getting to the customer, Amazon's process starts with the customer and works back to the business. Not only does Working Backwards create a better customer experience, but it's also more efficient because the tech team only builds what the customer truly needs.

Working Backwards is based on answering five key questions:

1. Who is the customer?
2. What is the customer problem or opportunity?
3. Is the most important customer benefit clear?
4. How do you know what customers need or want?
5. What does the customer experience look like?

At Amazon, this is typically done by creating a fictional Press Release with Frequently Asked Questions, called a PRFAQ. The press release defines the solution from the customer's point of view, while the FAQ section answers questions that customers and internal stakeholders might have, helping to flesh out the solution.

The Press Release

In a typical product development cycle, the press release comes at the very end. After the product is built, we step back and tell the customer, "Hey, world! We built this awesome thing, and you should love it!" The company tries to convince the customer to buy something it thought was a good idea. Instead of building something customers want, it's trying to convince them to buy what it built.

Working Backwards flips this process on its head. Instead of trying to sell a product after it's built, Amazon builds the product the customer wants. This ensures that the development process is always aligned with delivering value to the customer. By starting with the customer's needs and working backward, the team can focus on solving real problems and creating delightful experiences.

That's why Working Backwards begins with a fictional press release. At Amazon, the press release is created at the start of the development process, defining what the end customer experience should be. It serves as a north star, guiding the entire project and ensuring everyone involved is aligned with the ultimate goal of meeting customer needs. This approach keeps the team focused and motivated, knowing they're working toward a well-defined, customer-centric outcome.

The press release offers a clear view of what the customer will experience at the end of the process. It answers key questions like, "Who is the customer using this product?", "What's the key pain point the customer has?", and "How will this product delight them?" Once the product is built, the final press release should be very similar to the fictional one from the PRFAQ, just with real dates and customer quotes. They are long-term, values-based documents. In fact, the original press release for the Amazon Kindle holds up remarkably well even today.

The Structure of a Press Release

- **Heading/Subheading:** Name the product or feature in a way that will resonate with customers. Add a sentence describing the primary benefit.
- **Summary Paragraph:** This is your executive summary, providing a brief overview of the product and its key benefit. It tells the reader what the product is and why they should care. The format should look like a real press release, including a city, media outlet, and expected launch date.
- **Problem Paragraph:** Define the customer's problem from their perspective. Focus on the pain point the product is designed to address, showing that you truly understand your customer's challenges. This is the most important part of the Working Backwards process.
- **Solution Paragraph(s):** Once the customer's pain is clear, describe how you'll solve it. Highlight the specific features of the product that will address their problems.
- **Quotes and Getting Started:** Make it feel more real by adding quotes. Start with a quote from a senior leader sponsoring the project, helping them understand and support the product. Follow up with a fictional customer quote to illustrate the product's value. Use this fictional customer to reflect the problem

and solution you've built around them. End with a simple call to action, like how to purchase or where to get more information.

Frequently Asked Questions (FAQ)

Once you've built the press release and shared it internally, the questions will start rolling in. These typically fall into two categories: customer questions and internal questions.

Customer questions often make their way into the final FAQ, covering things like:

- "How much will it cost?"
- "Where can I buy it?"
- "How does it work?"

These questions help give customers the confidence they need to make a purchase.

Internal stakeholders will ask a different set of questions, such as:

- "How big is the market for this product?"
- "What will it cost to produce?"
- "What kind of revenue can we expect?"

These are the questions that help the internal team determine if the product makes business sense.

You won't cover every question, and the answers will inevitably change. But a solid FAQ forces you to think ahead and put together a strong strategy. As my friend Neil, an entrepreneurial CEO, once said, "If I can create a plan that shows this will work, it'll work 50% of the time. If I can't create a plan that shows it will work, it'll work 0% of the time." Surprises will happen, but by tackling key questions early, you can guide the development process and ensure you're focusing on the most critical issues right from the start.

Focusing on What Matters

This may seem like a lot of work—and it is. All of this happens before the project even starts, and naturally, everyone just wants to start building. During a town hall, someone asked Jeff, "Do we really have to do this for every project?" His response was, "If you know something better, use it. But try this method for a while."

The reason is simple: you need to clearly define the problem before you start building. Why begin development without knowing the customer experience you're aiming for? Why build something before you've answered the key questions customers will have? The Working Backwards process does the hard work up front, when changes are easiest and least costly. Waiting to make adjustments later becomes expensive, and those changes are less likely to happen. This process ensures you're building with purpose, not guessing at what the customer might want.

It's like writing user documentation. In traditional software development, user documentation is often an afterthought. It's not as interesting as building and designing new features. Software teams joke, "User documentation is very important. It's the first thing a customer sees. That's why it's given to the most junior member of the team." But in the Working Backwards process, this is flipped on its head. Teams write the user documentation before a single line of code is written, because if it doesn't affect the user, it's not worth building.

Example: The Amazon Kindle Press Release[19]

One of the early successes of the Working Backwards process was the Amazon Kindle. Consider what a challenge it was—creating not just a new product, but an entirely new business from the ground up. There were countless decisions to make, from hardware design to content

---

[19] Amazon, "Introducing Amazon Kindle," Amazon Press Center, November 19, 2007, https://press.aboutamazon.com/2007/11/introducing-amazon-kindle.

partnerships, and they all needed to line up. By starting with the customer's needs and working backward, the team was able to shape the Kindle in a way that felt intuitive, solving problems readers didn't even realize they had.

Here are the key components of the original Kindle press release. While it doesn't include the fictional elements—like the imagined quote and problem statement—from the Working Backwards draft, it remains substantively the same. Nearly two decades later, it still presents a compelling vision for what the Kindle set out to achieve.

- **Heading:** Introducing Amazon Kindle
- **Subheading:** Revolutionary Portable Reader Lets Customers Wirelessly Download Books in Less Than a Minute and Automatically Receive Newspapers, Magazines and Blogs
- **Summary Paragraph:** Amazon.com today introduced Amazon Kindle, a revolutionary portable reader that wirelessly downloads books, blogs, magazines, and newspapers to a crisp, high-resolution electronic paper display that looks and reads like real paper, even in bright sunlight. More than 90,000 books are now available in the Kindle Store, including 101 of 112 current New York Times bestsellers and new releases, which are $9.99, unless marked otherwise.
- **Quote:** "We've been working on Kindle for more than three years. Our top design objective was for Kindle to disappear in your hands—to get out of the way—so you can enjoy your reading," said Jeff Bezos, Amazon.com Founder and CEO. "We also wanted to go beyond the physical book. Kindle is wireless, so whether you're lying in bed or riding a train, you can think of a book, and have it in less than 60 seconds."

The final press release included these key elements—questions that the Working Backwards draft would also need to address, though it might explore ranges of possibilities rather than settling on final answers:

- **How much does it weigh?** 10.3 oz
- **What selection is available?** The Kindle Store currently offers more than 90,000 books, as well as hundreds of newspapers, magazines, and blogs.
- **How many books can it hold?** Built-in memory stores more than 200 titles, and hundreds more with an optional SD memory card.

By drafting the press release before the product was built, the Kindle team kept their focus sharp—anchored on what would matter most to the customer. The product evolved, as all great products do, but its core promise remained the same: a small, portable reader that lets people download and enjoy books from anywhere.

Understanding Your Customer

What's the best product to build? Naturally, we want a lot of customers, so it might seem like building something that appeals to everyone is the answer. But the problem is that different people want different things.

In the early days of desktop software, companies tried to compete with Microsoft Word by offering a minimal set of features. They assumed that by including just the 80% of features most users needed, they could cut out the rest. One of the features they often skipped was word count. Very few people need word count, but journalists and reviewers do. When these products were reviewed, journalists invariably pointed out how they didn't meet the needs of serious writers. "They didn't even have word count!" That feature might only matter to 1% of users, but it's an important 1%.[20]

This problem isn't limited to software development. Venture capitalist Ben Horowitz learned this firsthand when he was discussing interview techniques with top salesperson Mark Cranney. Expecting a

---

[20] Joel Spolsky, "Strategy Letter IV: Bloatware and the 80/20 Myth," Joel on Software, March 23, 2001, https://www.joelonsoftware.com/2001/03/23/strategy-letter-iv-bloatware-and-the-8020-myth/.

thoughtful case study question, Ben asked, "What's the best interview question you ever got?"

Cranney answered immediately: "What would you do if I punched you in the face right now?"

"What?!" Ben responded. "He wanted to know what you would do if he punched you in the face? That's crazy. What did you say?"

"I asked him, 'Are you testing my intelligence or my courage?' And the interviewer said, 'Both.' So I said, 'Well, you'd better knock me out.' He said, 'You're hired.'"[21]

I would never ask that kind of question in an interview, but I'm looking for different qualities when I hire someone. Cranney realized he was being tested on his competitiveness, his ability to handle pressure, and his understanding of the intent behind the question. These are key traits for a salesperson but not for a product manager.

Doing a Great Job: Obsess Over Your Customer

Once you've identified your customer, the next step is understanding their problem—not what you think their problem is, not what you're good at solving, but their actual problem. This is where most companies go wrong. They solve the problems they're good at addressing, not necessarily the ones their customers need solved. It's the classic case of a man with a hammer seeing everything as a nail.

People with great tools really want to use them—like surgeons. Surgeons have been trained to cut, so that's often their go-to move. But no matter how skilled a surgeon is in the operating room, they can't be truly great without listening to their patients and understanding what they really need. A great surgeon sits with the patient, takes the time to understand and define the problem, and then decides how to fix it. The goal isn't just a successful surgery—it's helping the patient live a better life.

---

[21] Ben Horowitz, *What You Do Is Who You Are* (New York: Harper Business, Kindle Edition, 2019), pp. 193–4.

In his book *Being Mortal*, Atul Gawande tells the story of Joseph Lazaroff. Gawande was a surgical resident when he met Lazaroff, a man in his sixties suffering from incurable prostate cancer. His lower body was filling with fluid, and he had recently woken up with a paralyzed leg. Surgery was an option, but it would only slow the spread of the paralysis and extend his life by a few months. It was a very risky operation, with serious complications, including death. But Lazaroff pleaded with Gawande, "Don't you give up on me. You give me every chance I've got."

Early in his career, Gawande did what he was trained to do. Using all the skills he'd learned, he performed an intricate operation, removing the tumor around Lazaroff's spine. The surgery was a success, but Lazaroff died two weeks later from complications. Gawande had solved the wrong problem. He had been fighting against disease and death when instead, he should have been fighting to give the patient the best possible life, given the circumstances.[22]

Doctors can be narrowly focused even when it's not life or death. My friend Richard told me about how his fellow physicians sometimes unintentionally harm their patients—simply by seeking more information.

Here's a typical situation: A patient comes to see Richard with an issue that hasn't been diagnosed by the usual tests. So, the doctor starts digging, ordering a huge battery of tests to figure out what's wrong. On the surface, this seems like a logical approach. But here's what makes Richard stand out. He says, "If you are having some troubling medical symptoms, buy life insurance. Once you go to the doctor and start getting tests, they might find something really bad. After that, it could become very expensive or even impossible to buy life insurance—and that's exactly when you'd need it the most."

---

[22] Paul Kalanithi writes about this in his book *When Breath Becomes Air*. Kalanithi was a promising young neurosurgeon when he was diagnosed with pancreatic cancer in his mid-thirties. He wrote of the difficult process of determining how to wring the best moments out of a fading life, and how he traded his physical health to ensure a sharp brain for as long as possible.

Focusing on the customer isn't just relevant for white-collar workers like product managers and doctors. I was speaking with the building manager of a high-end New York property, and he shared a surprising quality he looks for in doormen. Buildings in New York may have hundreds of residents. Luxury buildings have a doorman who protects residents and lets them in without a key.

While being able to handle intruders is important, the top priority is their ability to recognize residents. This allows them to seamlessly keep out trespassers—those who don't belong—while warmly welcoming those who do.

Final Thoughts

At the heart of everything Amazon does is a relentless focus on the customer. Amazon's culture isn't just about solving problems—it's about solving the right problems for the customer. This obsession with understanding and addressing customer needs is woven into the company's DNA, from product development to everyday interactions.

The company has scaled this principle from Jeff's original insight to the entire organization through the Working Backwards process. Instead of creating products that Amazon thinks people want and then pushing them onto customers, the company starts with the customer, defining the end experience before building anything. By writing a press release at the beginning of the process, Amazon sets a clear vision for success, rooted in what will truly delight customers.

The principle of Customer Obsession extends far beyond Amazon. I've seen it in industries as diverse as healthcare and security. By understanding who the customer is and what they really need, you can develop far better solutions. This is the essence of Customer Obsession, and it works in any field.

On a personal level, I applied this with a mentee. When I taught her to think using Working Backwards, she completely transformed her career.

About a decade ago, I was speaking with Debbie, a promising new recruit. She was looking for career advice, asking me questions like, "What was the most important factor in your success?" and "What's the best advice you ever received?" After answering, I told her, "These are great questions, but they're not the right questions."

She looked confused and asked, "What do you mean?"

"You're asking what I've done to be successful. You should be thinking about what you want to do."

Debbie told me that she wanted to be a product manager, like me. She was going to get a master's degree in information science and then get her MBA—at least, that's what her parents were pushing. I told her to take a step back and really think about what she wanted, rather than what others expected of her, and that we could talk again once she'd had some time to reflect.

We talked a lot over the next few years. She realized that software development wasn't her passion after all. Instead of jumping into MBA programs right away, she accepted a position at a prestigious consulting firm, where they offered to sponsor her MBA in the future.

By applying the principle of Working Backwards, Debbie shifted her focus from what others expected of her to what she truly wanted. It was a big change, but it aligned with her real passion. Today, she's on the path to becoming a doctor. And it all started with solving the real problem—not what others thought she should do, but what she actually wanted.

How to Obsess Over Yourself

If you want to be like Debbie, try some of the exercises below that I've used with my mentees over the years. These will help you define your goals, understand your strengths, and align your career with what truly matters to you.

- **How to find your perfect job?** My friends created *Never Search Alone*, a free organization that connects jobseekers with others on the same path. Together, you work to identify what you really want to do and how to get there. They use a great tool called the *Mnookin 2-Pager* to help clarify goals, which I've found very useful.

- **Where have you been successful?** It's a broad question, but a good place to start is by reflecting on what you've done well and what you've enjoyed. When I mentor people, I often use the *Seven Stories Exercise* from The Five O'Clock Club, which provides a strong framework for capturing this.

- **What do you like to do?** You'll find work most fulfilling if you can be your authentic self. You can do this by knowing what skills and qualities you value most. By looking at what you value most (and least), you'll see what kinds of jobs will make you most fulfilled. I like the *VIA Character Strengths Survey* to help define values.

- **Write your future story.** Just like Amazon writes a press release before developing a product, write your own life story. Imagine yourself 5 or 10 years from now—what do you want to have accomplished? What does success look like for you? If you can clearly define the outcome, it becomes much easier to work toward it.

LP #2

OWNERSHIP

Leaders are owners. They think long term and don't sacrifice long-term value for short-term results. They act on behalf of the entire company, beyond just their own team. They never say "that's not my job."

Amazon has a decent chance of being the last place to buy CDs. The business will be high-margin but small. You'll be able to charge a premium for CDs since they'll be hard to find.

—Steve Jobs to Jeff, *Working Backwards*, pp. 164–65.

In 2003, Steve Jobs invited Jeff and a few other senior Amazon executives to Apple's headquarters in Cupertino. The iPod was becoming the dominant music player in the world, but Steve wouldn't open it up to the Windows ecosystem. It was a market-limiting move for the iPod, used to drive Mac sales.

At this meeting, Steve revealed that Apple would make iTunes available for Windows, allowing Apple to sell digital music to anyone. That's when Steve threw down the gauntlet. He told Jeff, "Amazon has a decent chance of being the last place to buy CDs. The business will be high-margin but small. You'll be able to charge a premium for CDs since they'll be hard to find."

The digitization of media was a huge threat to Amazon's core business of selling and shipping physical books and CDs. So in January 2004, Jeff started Amazon's digital media business. But how he did it was important. He didn't jump in and try to create a better version of iTunes. In fact, he didn't have any specific product or business in mind. His first decision was to appoint an owner to the project.

He promoted Steve Kessel to the job. Previously, Steve was Amazon's head of books, video, and music. Steve became a "single-threaded leader," an Amazonian term for a leader who is 100% dedicated to and accountable for a specific project or initiative.

He was able to drive this new business unencumbered by the rest of Amazon's operations. It also allowed the physical media business to thrive without worrying about digital. He was able to look at the market and create an entirely new business for Amazon: Kindle. By splitting responsibilities between today and the future, Amazon was able to focus on being the world's best physical book e-commerce site while eBooks eventually eclipsed physical ones.

Single-Threaded Architecture

The idea of the single-threaded leader comes from Amazon's computer systems. In the early 2000s, Amazon.com ran as a single massive program named Obidos, named after a village on the fastest part of the Amazon River. Obidos was built to run fast. Amazon also had a single database that ran the entire company—an Oracle database called "acb" (short for amazon.com books).

As Amazon grew, Obidos and acb became two huge bottlenecks for change. The speed of innovation slowed down, and developer frustration skyrocketed. Also, having one single application meant that a junior developer could crash the entire website—which happened more than once.

At the time, building a system this way was fairly common. Most businesses started with one big application because it's the fastest, easiest, and most efficient system to develop. In the early days of enterprise computing, computers were giant, room-sized machines. They spoke an arcane language and used specially formulated punch cards. Computing power was the scarcest resource in the company. A mistake in a punch card could cause the business to waste thousands of dollars in lost processing time.

To run these machines at peak efficiency, a cadre of high priests of computing emerged, tending to their every need. Much like ancient gods, these priests' main goal was to keep the machines happy with their daily supply of punch cards.

Over time, things changed. Computers got cheaper, shrinking in size and embedding themselves into our personal and work lives. But software continued to be built as large applications in most places. As software expanded, the speed of software creation started to slow down.[23]

In theory, adding more people to a project should speed things up when progress lags. But in reality, it doesn't work that way. Fred Brooks wrote about this problem in his book *The Mythical Man-Month*. Brooks explained that while adding another person to the project increases work capacity, it also increases communication complexity. Things slow down because the number of connections grows much faster than the additional output of the new worker.[24] Here's a graph

---

[23] Steve Jobs gave a fascinating talk on "Fractional Horsepower Computing" at the Aspen Institute where he compared computers in 1984 with the power of motors. Jobs talked about how early factories were set up next to a water source that drove a giant power source. This meant that every motor was coupled to that one drive train. Once electric motors were invented, factories could become much more efficient by putting small amounts of power where they needed to be. Before that, if you wanted a sewing machine, it needed to be huge, like a weaving loom. Now you could have something the size of a sewing machine. https://www.youtube.com/watch?v=Vd-Fpm4mCdqs

[24] Brooks's solution to the problem was to have a small number of programmers who were sup-

showing how connections multiply as more people join. You can see
the exponential rise in communication complexity, which explains
why bigger teams can be harder to coordinate.

Growth of Communication Complexity

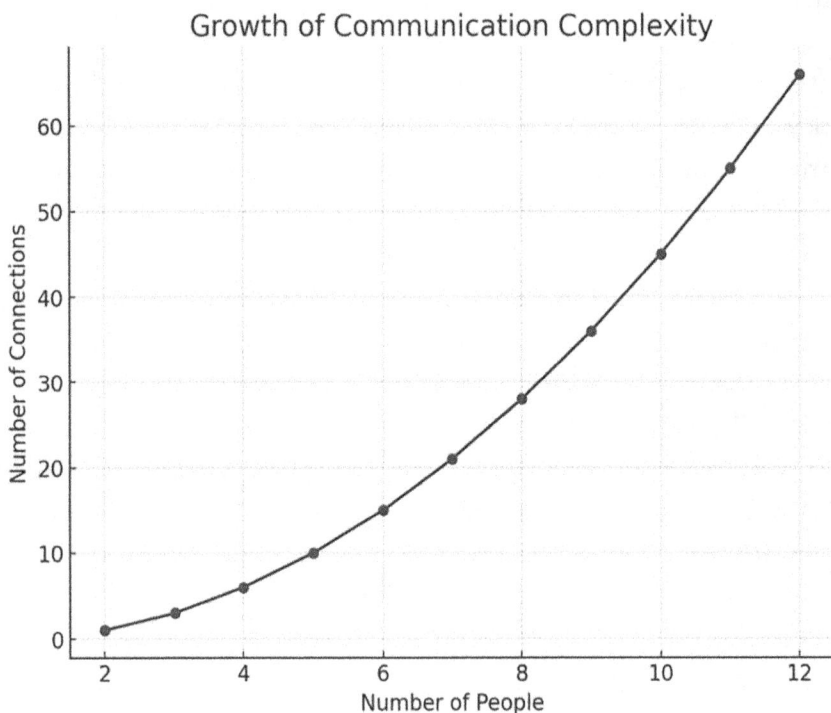

Growth of Connections vs. People

Looking at the graph, it's clear why Amazon prefers to keep teams
small. But how small? Amazon has a term for it—Two-Pizza Teams.
It was an informal way to define the ideal size: small enough to be
fed with two large pizzas. In practice, that means about seven to nine
people—big enough for collaboration and some scale but still small
enough to keep communication smooth and decisions fast.

But can a company as massive as Amazon really run on teams that
size? That's where microservices come in.

---

ported by a phalanx of support staff to make them as productive as possible. This made each individual programmer more productive, which reduced the number of communication channels.

Microservices

Today, Amazon builds at lightning-fast speeds even though it has tens of thousands of programmers making millions of tiny changes a year. Amazon Web Services is robust enough to run Capital One's entire bank, McDonald's ordering system, and the State of California's 911 system.

How was Amazon able to scale so well? It broke up the different pieces of the system and separated them. Each piece focused on one individual job. This approach is a bit counterintuitive. You'd think that the best way to build things would be for everyone to communicate with each other. Communication is a good thing, right? However, as we saw above, excessive communication is a productivity killer. With microservices, communication takes on a very different form.

The original Obidos system had too many cooks in the kitchen. Imagine starting a food service business called "Amazon Meals." Initially, you have one kitchen shipping orders nationwide, and you have one or two chefs. But as the company grows, you might have a hundred chefs in your kitchen.

If the Latin chef decides to order a new type of tomato for Spanish tortilla soup, it might make for a fantastic soup but accidentally ruin the pizza recipe, which relied on the old tomatoes. Well, we can't have that happen! So what can we do?

To avoid constant recipe conflicts, you could introduce a process that prioritizes and approves changes. But this approach would be slow and frustrating for everyone. This is what was happening in Obidos.

A better solution is to break things up—give each team its own kitchen, so to speak. Instead of one chaotic space where every chef's decision ripples through the entire operation, you divide responsibilities: a sauces team, a produce team, a bakery. Each group publishes its offerings like a menu, making it clear what's available without endless coordination. If the soup chef wants a new type of tomato, they don't just swap it in and risk ruining the pizza—they request it from the produce team, who can decide if it makes sense to add it. No one needs to micromanage anyone else; they just rely on the ingredients that are officially on the menu. The result? Each team moves faster, innovation happens without bottlenecks, and best of all—no recipes get ruined in the process.[25]

This is how microservices work. Instead of one sprawling system where every change risks breaking something else, you have independent services, each focused on a specific job. Like our kitchen teams, these services don't need to know each other's inner workings—they just rely on what's on the menu. In tech, this "menu" is the API (Application Programming Interface), a structured way for services to communicate without interference. Each Amazon "two-pizza team" owns its own microservice, making updates without disrupting others.[26]

The Job of a Minimum Viable Product (MVP)

Now we know *how* to build, but *what* should we build?

Most companies rely on the HiPPO (Highest Paid Person's Opinion) method for product strategy—the boss decides, so they must know what to build. The problem? While great leaders have plenty of skills, predicting what customers will want isn't usually one of them. In fact, no one can perfectly predict future customer demand.

That's where the Minimum Viable Product (MVP) comes in. An MVP is the simplest thing you build to solve a customer's problem. It's not a finished product, but an experiment. It's about making a specific promise to customers and delivering on it. You have to release

---

[25] This process comes with its own challenges. Every station, from the grill to the bakery, needs to be solid because the whole operation depends on it. At the same time, each station must protect itself from others taking more than their share. If the bakery suddenly floods the ovens with bread, there's no room for anything else.

[26] If you want to understand Amazon's microservices architecture, it's worth reading this famous post. In 2011, a former Amazon engineer accidentally made an internal rant public on Google+, comparing how Amazon operated to Google—and not in Google's favor. In the process, he revealed how Amazon had embraced microservices long before it was mainstream, breaking systems into independent services that could scale efficiently. This shift not only helped Amazon grow but also laid the foundation for what would become AWS. https://gist.github.com/chitchcock/1281611

your MVP before every "i" is dotted and "t" is crossed. Reid Hoffman, the founder of LinkedIn, says, "If you're not embarrassed by the first version of your product, you've launched too late."

That word "embarrassed" is often misunderstood. When you launch with an MVP, you won't have all the answers, so looking back, you should have a moment of *"If only I had known..."* But that's exactly the point—you *didn't* know. Embarrassed doesn't mean careless, reckless, or unprepared. It doesn't mean launching something broken, ignoring obvious problems, or rushing forward without a plan. It just means you moved fast enough to start learning.[27]

Why do we have to build an MVP? Why can't we just build everything at once?

Imagine you have a birthday party coming up. You worked late last night and don't have enough time to bake the cake you promised a friend for her birthday. You have exactly half the time you originally planned. You can either:

1. Bake half of the cake.
2. Bake a half-baked cake.

Baking half a cake is clearly the superior solution. You end up with a finished product that works for the birthday party. People get smaller slices, but they still get cake. They eat less and are excited for more.

Baking a half-baked cake doesn't work. No one wants to eat a gooey mess. Yes, you "tried your hardest" to meet the original spec, but that doesn't make it taste better.

The Zappos MVP

In 1999, Nick Swinmurn went to a San Francisco mall looking for brown Airwalk Desert Chukka boots. He spent hours searching. One

---

[27] Reid Hoffman, "If There Aren't Any Typos in This Essay, We Launched Too Late!", LinkedIn Pulse, March 29, 2017, https://www.linkedin.com/pulse/arent-any-typos-essay-we-launched-too-late-reid-hoffman/.

store had the right color but the wrong size. Another had the right size but the wrong color. Frustrated, he thought he could do better with an online shoe catalog.

He could have pitched his business plan to venture capitalists, built a warehouse, and stocked inventory. Instead, he built a simple website called ShoeSite, where customers could browse and order shoes.

But this was just an MVP—a test. When people ordered shoes from the site, Nick would go to the mall, purchase the shoes, and ship them to the customer.

Zappos's first MVP was so simple that, looking back, it's almost embarrassing. Nick didn't even buy the shoes at a discount. But the MVP provided enormous benefits. It proved that customers were excited to buy shoes online. It also revealed an important pain point: if customers can't try shoes on before buying, they need an exceptional return policy. That insight led to Zappos offering free two-way shipping and a 365-day return window.

Keeping Small Promises

A few years ago, I was working with a junior team of developers on an MVP. Our customers had five key use cases they cared about, and the team suggested tackling a little bit of each. "That way," they reasoned, "we can provide a full-service solution and make all of our customers happy."

I pushed back. "The problem isn't just what we build—it's how people experience it. If we release five mediocre features, people will say, 'This doesn't work that well.' But if we build one feature that truly solves a problem, people will come back for more."

I told them about an old Head & Shoulders commercial from the 1980s. In the ad, a man is on a date with a beautiful woman. Everything is going great until she notices dandruff on his shoulders. Suddenly, she's completely turned off. The tagline? "You never get a second chance to make a first impression."

That's how products work, too. If your first version is underwhelming, customers won't give it another shot. It's fine if the product needs tweaks and improvements, but it can't be something people immediately dismiss. You don't need to solve everything—you just need to make one solid promise and deliver on it.

Later that evening, we had a big happy hour with some senior leaders. One of the junior team members told me she had met a few executives but not as many as she had hoped. She looked disappointed. "I thought you said you never get a second chance to make a first impression."

I laughed. "Your happy-hour performance was like an MVP. The people that you did talk to saw that you were curious, engaged, and eager to learn. Even those who didn't speak with you probably noticed that energy. You can build on those conversations next time. If you had rushed around trying to talk to everyone, you wouldn't have given anyone your full attention, and that could have come across as scattered or insincere. Besides, at a happy hour, as long as you don't annoy people or get horribly drunk, you've done just fine."

Final Thoughts

While we like to pretend we can do everything, we accomplish much more when we focus. This is why Amazon has single-threaded leaders and focused Two-Pizza Teams. By organizing efforts this way, teams can be clear on their commitments and more likely to deliver for internal and external stakeholders.

Creating a firm sense of ownership allows the company to scale like few organizations ever have. It helps teams focus on solving a specific customer problem. Just like building an MVP:

1. You're never going to do everything. There will always be compromises.

2. Instead of compromising on quality, cost, or speed, reduce your scope—focus on the customer.
3. Focus on customer needs, not what you want to build.
4. Make promises to customers and keep them.

I'll end with a story about focus and owning one thing at a time.

We like to think we're great at multitasking, but research shows we perform better when we give something our full attention. It's tempting to believe that if doing one thing is good, doing two at once must be even better. But some things just don't mix well.

Take tea, for example. Some days, I want iced tea because it cools me down. Other days, I want hot tea because it heats me up. But I've never once thought, "What I could really go for right now is a big cup of lukewarm tea."

My friend Marc has a great trick for single-tasking. The last time we met for lunch, he pulled out his phone as soon as we sat down. "I hope you don't mind," he said, "but I'm setting a timer. That way, I don't have to check my watch or worry about time. I can just focus on our conversation."

Instead of splitting his attention between me and the clock, he was fully present—like it was just the two of us in the room, no ticking third wheel.

LP #3

INVENT AND SIMPLIFY

Leaders expect and require innovation and invention from their teams and always find ways to simplify. They are externally aware, look for new ideas from everywhere, and are not limited by "not invented here." As we do new things, we accept that we may be misunderstood for long periods of time.

At first I had to go deliver the books to the post office myself. I don't still deliver, but I was doing that for years. In the first month I was packing boxes on my hands and knees on the hard cement floors. I said to the person kneeling next to me, "You know, we need kneepads because this is killing my knees," and he said, "What we need are packing tables"—the most brilliant idea I'd ever heard. The next day I went and bought packing tables and doubled our productivity.

—Jeff[28]

Invention fills the gap between a customer's needs and a company's ability to meet them. We've already covered what the customer

---

[28] This story is an example of a "Jeffism." Brad Stone uses it in *The Everything Store*, and you can also find it in Jeff's Economic Club interview.

wants **(LP #1: Customer Obsession)** and who is responsible **(LP #2: Ownership)**. Now it's time to start delivering.

Invent and Simplify requires a new way of thinking. It's about understanding what customers need and making their lives a little bit better—or maybe just a little bit less bad.

I'm a product manager. Most people think a product manager's job is to build new and innovative products, but that's not quite right. It's not about making a cool new product; it's about making a product that solves real problems for customers. Let's take a look at the Amazon Alexa virtual assistant. Alexa isn't just a cool new technology; it improves the way my family works together.

Innovating in the Living Room

The living room of the twenty-first century didn't immediately call out for innovation. It was a room to talk to each other and watch TV—a quiet space to sit. But then cell phones and smart technology arrived.

Family members would pause conversations to check the weather or set a kitchen timer. As homes became smarter, they pulled out their phones to adjust the thermostat or turn on the lights. But the moment they opened their phones, a notification would pop up—an email, a text, or a new social media update—leading them down a digital rat hole. The phone was a way of taking a person out of the group and moving them somewhere else.

Phones became a means of teleporting people into their own world. To make sure our family paid attention to each other, we made a rule: No phones at the dinner table.

Alexa, on the other hand, feels like part of the group. Instead of one person pulling out their phone and tuning out, anyone can just ask, "Alexa, what's the weather?" and keep the conversation going. It's a small shift, but it keeps everyone present instead of disappearing into their screens.

Alexa has almost become a member of the family. I saw this firsthand as my kids grew up. Alongside traditional developmental milestones like learning to sit up, learning to talk, and learning to count, there was now "learning to speak well enough for Alexa to understand you."

One of my friends told me about her two-year-old screaming, "Lex-ah, Lex-ah, Lex-ah, you listen me!" when she wasn't understood. A year later, she ran into her room, excited to share great news with her virtual friend: "Alexa! Alexa! I got a big girl bed."[29]

Amazon starts with the customer rather than with technology. It's about focusing on the "home" in "smart home" rather than the "smart." Technically, Alexa is a series of complex AI systems, speech recognition, and speech-to-text processing. But these technologies come together in a way that simplifies life.

Simple Is Harder Than You Think

We like to assume some things are so simple they don't need a second thought. People say, "It should be as easy as tying your shoes." But as Terry Moore pointed out in his TED Talk, most people have been tying their shoes wrong for years without realizing it.[30]

I was reminded of this one day at work when I struggled with something that should require no thought at all—opening a door. I was at work one day with my friend Mitchell. I'd only been at this job a few weeks. When he was getting up to leave, I started pulling and pushing the door, unsure of which way it opened.

Mitchell said, "I guess you haven't figured out how the doors work yet."

"No," I answered, "and I don't intend to. It's the door's fault. The doors are designed wrong."

---

[29] Because Alexa is becoming so much like a person and kids interact with it that way, we had a rule, since abandoned, of saying "please" and "thank you" to Alexa. The personification of Alexa and other assistants has some interesting human implications, which have become even more present with the prevalence of ChatGPT.

[30] Terry Moore, "How to Tie Your Shoes," TED video, February 2005, https://www.ted.com/talks/terry_moore_how_to_tie_your_shoes.

Mitchell doubled over in laughter. "What do you mean, 'It's the door's fault'?"

"Look at this door," I said. "The handle is a big vertical bar—on both sides. The architect thought this would look nice and create symmetry. However, a vertical bar signals that you should *pull* the door. To signify that you should *push* a door, you'd want a flat panel or something similar."

Mitchell laughed; he'd never realized a door could be so complicated.

These types of hard-to-open doors are common. They even have a name: "Norman Doors" after Don Norman, the author of the foundational design book *The Design of Everyday Things*. He points out that many everyday products are designed with an eye toward beauty and fashion, without considering how people actually use them. Instead of helping us get things done, these tools fight us—a phenomenon Norman labeled the "Psychopathology of Everyday Things."

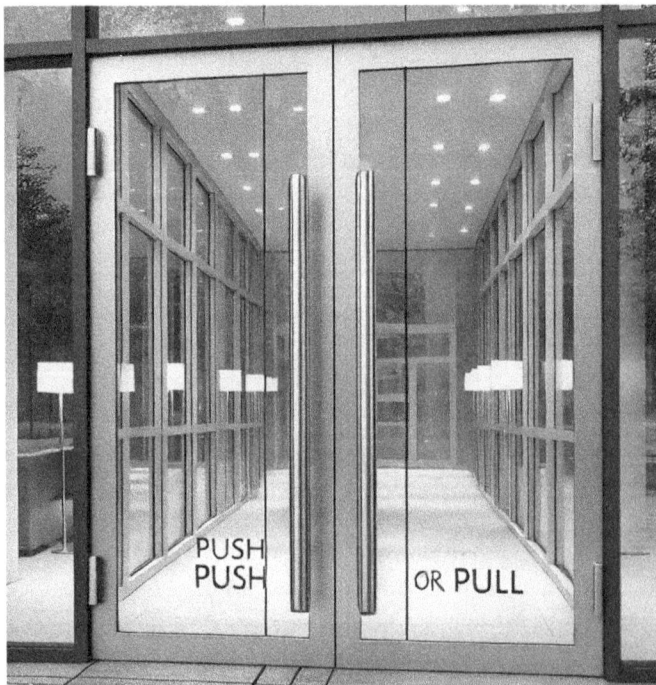

Norman Door

The Dangers of Complexity

When we add features to a new product and make it more complicated to use, we might think it's no big deal. It's easy to argue that some people like lots of features and those who don't will just ignore them. But there can be a real danger in making something complicated or confusing. Poorly designed products can cause serious problems—and even death.

A few years ago, Tylenol faced a serious and far-reaching problem. When I think of a Tylenol crisis, I think of product tampering—of someone opening bottles and slipping in poison.[31] But this was different. There was no outside culprit—just a well-intentioned but deeply flawed design decision that made it far too easy for parents and even doctors to give infants the wrong dose. And because the risk wasn't obvious, the danger kept spreading, unnoticed until real harm had already been done.

The company was trying to be helpful. They created two different versions of children's Tylenol—one for children and one for infants. This allowed parents to give the medicine to young children using a small liquid cup, while treating infants with an even smaller eye dropper. But to make this work, the infant formula was about three times more concentrated than the children's version. To see how complicated this was, take a look at "Old Tylenol Infant Drops" and "Tylenol Infant/Children's Liquid" below:

---

[31] This happened in 1982 when seven people in the Chicago area died after taking Tylenol capsules laced with cyanide, leading to a nationwide recall, stricter packaging regulations, and one of the most infamous unsolved crimes in U.S. history.

Acetaminophen (Tylenol) Dosing Chart

May give acetaminophen dose every 4–8 hours; no more than 5 doses in 24 hours

Weight	Tylenol Milligram Dosage	Old Tylenol Infant Drops 80mg/0.8ml	Tylenol Infant/Children's liquid 160mg/5ml	Tylenol Chewables 80mg each	Tylenol Junior 160mg each
6 - 8 lbs	40 mg	½ dropper (0.4 ml)	¼ tsp (1.25 ml)	N/A	N/A
9 - 11 lbs	60 mg	¾ dropper (0.6 ml)	⅓ tsp (1.875 ml)	N/A	N/A
12 - 17 lbs	80 mg	1 dropper (0.8 ml)	½ tsp (2.5 ml)	N/A	N/A
18 - 23 lbs	120 mg	1 ½ dropper (1.2 ml)	¾ tsp (3.75 ml)	N/A	N/A
24 - 35 lbs	160 mg	2 droppers	1 tsp (5 ml)	2 tablets	1 tablet
36 - 47 lbs	240 mg	3 droppers	1 ½ tsp (7.5 ml)	3 tablets	1 ½ tablet
48 - 59 lbs	320 mg	N/A	2 tsp (10 ml)	4 tablets	2 tablets
60 - 71 lbs	400 mg	N/A	2 ½ tsp (12.5 ml)	5 tablets	2 ½ tablets
72 - 95 lbs	500 mg	N/A	3 tsp (15 ml)	6 tablets	3 tablets

It seemed like a great product. "Have an infant that can't use a cup? We have the product for you! It's a super concentrated Tylenol you can dispense from an eye dropper." But here's how it worked for us. One day in 2010, we took my infant son to the doctor. He had a fever, and the nurse gave him Infant Tylenol. But instead of giving him the infant dosage, she told us to give him the much larger dosage meant for Children's Tylenol.

When I pointed this out, she said, "That can't be right—it's so tiny." But after checking with the doctor, she realized she had been giving infants three times the correct dose. She had advised me to overdose my son to a dangerous degree.

Clearly, this was a *failure* on the company's part:

1. If a trained nurse found the system confusing, it was too complicated for the average parent.
2. Tylenol should have tested the packaging and dosing instructions with real users.

While this was dangerous—possibly even lethal—I didn't make a big deal about it at the doctor's office. At the time, I didn't fully grasp the risk. Looking back, I should have pushed for the doctors to notify patients and make sure they were taking the right dose. But in the moment, it felt like just another avoidable human error. If everyone simply read the instructions correctly, everything would be fine—or so I thought.

Building Products for Real People

Tylenol created a product that was too easy to get wrong, and the consequences were deadly for some parents.[32] This kind of mistake happens when product designers and engineers focus too much on the product and too little on how people actually use it. And when things go wrong, the blame usually falls on the user. They call it "human error," as if the flaw isn't baked into the design itself. The most narcissistic designers will even go further, saying, "The product would work perfectly if we could just take the people out of the equation."

Software developers often fall into this trap as well. A team at Oregon State University studied why so much software feels like it wasn't made for real people.[33] They use personas (technical speak for example people) to explain the disconnect between the way developers think and the way users think. Let's meet Tim and Abby.

Tim, the software developer, loves technology. He enjoys figuring things out, clicking around new interfaces, and discovering hidden features. He feels confident in his tech skills, so if something doesn't work and he can't fix it, he blames the technology. When he can't find

---

[32] Tylenol is a particularly dangerous drug. Because it is easier on the stomach than aspirin, it's marketed as a safer alternative. This American Life and ProPublica produced an exposé of how acetaminophen is secretly the most dangerous over-the-counter drug in the country, killing 150 people a year. Tylenol eventually stopped selling the infant formula due to this confusion. https://www.thisamericanlife.org/505/use-only-as-directed

[33] These examples are based on an Oregon State University tool called GenderMag that helps software designers focus on creating more inclusive tools. Originally, it was developed to help software engineers understand differences in how male and female customers interact with technology, but it works well for general users as well. While it may seem stereotypical, the personas are based on copious amounts of research. For more information, check out https://gendermag.org/.

an answer, he asks Google. For Tim, technology is a game, and the fun is in solving it. So when he designs software, he builds for people like himself—comfortable exploring, unafraid to tinker, and happy to dig into the source code when needed.

But most people aren't like Tim. They are like Abby. Abby doesn't love technology for its own sake—she just wants to get something done. She has a goal, and software is just a tool to help her reach it. She doesn't click around just to see what happens. If she's not sure what a button does, she won't press it. If she can't find a feature quickly, she assumes it doesn't exist. When things don't work, she blames herself, not the technology. When she can't find an answer, she asks a friend. For Abby, software isn't fun—it's a means to an end.

The problem is that Tim assumes everyone thinks like Tim. He builds software that rewards exploration, expects technical confidence, and assumes users won't mind a little trial and error. But Abby doesn't work that way. If the interface is unclear, Abby gets frustrated. If an action feels risky, Abby hesitates. If she can't find what she needs quickly, she gives up—not because she's bad at technology, but because she doesn't have time to wrestle with it. Tim's software isn't necessarily bad, but it could be so much better.

Software developers need to think more about Abby—the people who don't geek out over technology for its own sake. Abby doesn't care about specs or features; she cares about what a product can do for her. As the famous Harvard marketing professor Ted Levitt once said, "People don't want to buy quarter-inch drills; they want quarter-inch holes."[34]

Abby has a job that needs to be done. She doesn't care how the problem is solved—only that it's solved easily. The value she gets isn't tied to the complexity of the solution. And that benefits users who are

---

[34] This is from the article "What Customers Want From Your Products" in *Harvard Business Review*, January 2006. Though the quote is commonly attributed to Levitt, he says that it comes from Leo McGinneva. This is an example of Stigler's Law of Eponymy: "No scientific discovery is named after its original discoverer." Stigler's law is, itself, an example of the law. Stigler, a professor of statistics, wrote that he learned the law from the sociologist Robert Merton.

like Tim too. He won't complain if things are better designed—if a feature is easier to find or an error message actually helps him.

Valuable Does Not Equal Hard

The best ideas often feel obvious in hindsight—so natural that people assume they were always there.

In the late 1990s, when I was studying computer science at Yale, Amazon patented "1-Click" technology, letting customers skip the shopping cart and buy instantly. At the time, it felt like Amazon was cashing in on common sense—it should have been free for everyone to use.

As technology guru Tim O'Reilly wrote at the time, "It's a classic example of the kind of software patent that would never be granted if the patent office had even the slightest clue about software... I'd be very surprised if there isn't a fair amount of prior art... There's nothing new in what you did."

Jeff pushed back. What looks like common sense in hindsight wasn't so obvious when it first came out. To settle the debate, he and Tim O'Reilly offered a $10,000 reward to anyone who could prove 1-Click wasn't new. They searched everywhere, expecting to find at least one forgotten implementation. But they didn't. How could this be?

Jeff explained that the 1-Click innovation was a change in the shopping model. Before 1-Click, people followed the real-world shopping process: go to the website, put things in a shopping cart, and check out. 1-Click removed the cart, allowing people to buy things instantly.

At the time, everyone was locked into the shopping cart metaphor because that's how stores work in the real world. But online, you don't need a shopping cart. If you want something, just buy it. Though it seems obvious in hindsight, no one had thought of it before.[35]

---

[35] Tim O'Reilly, *What's the Future and Why It's Up to Us* (New York: Harper Business, 2017), pp. 71–75.

Final Thoughts

As a product manager, my job is to create products on behalf of my customers. The phrase "on behalf of" is critical—I'm here to serve them, not the other way around. My goal is to give them superpowers, enabling them to accomplish things they couldn't do before. To achieve this, I must constantly simplify my products and make them easier to use. When a customer finds something too complicated, that's my fault, not theirs. And the only way to truly understand their struggles is to step outside my office (or call them up on Zoom) and watch them use the product. Observing real users in action often reveals pain points that no survey or analytics dashboard ever could.

The best solutions are often the simplest—ones laser-focused on solving the problem without adding anything extra. These solutions are often so self-evident in hindsight that they're overlooked or dismissed as trivial. Take 1-Click ordering. All it did was remove a few unnecessary steps from the checkout process, but that small shift completely changed how people shop online.

LP #4

ARE RIGHT, A LOT

Leaders are right a lot. They have strong judgment and good instincts. They seek diverse perspectives and work to disconfirm their beliefs.

I predict one day Amazon will fail. Amazon will go bankrupt. If you look at large companies, their lifespans tend to be 30-plus years, not a hundred-plus years.

—Jeff, Amazon All-Hands Meeting, November 2018

Each quarter, Jeff and the leadership team gathered employees for an All-Hands meeting streamed around the world. Often it was about the innovative work the company was doing, like the launch of Amazon Go convenience stores or the acquisition of Whole Foods. In November 2018, Jeff spoke about his concern for the company, warning that if we ever stopped thinking like a startup, Amazon would die. We'd heard it many times before,[36] so I didn't think much of it when he said, "I predict one day Amazon will fail. Amazon will go bankrupt."

---

[36] This was really just Jeff's normal warning about Day 2, like in his 2016 letter when he said, "Day 2 is stasis. Followed by irrelevance. Followed by excruciating, painful decline. Followed by death."

The press, however, had a different opinion. With hundreds of thousands of employees, someone was bound to talk. Some outlets ran with it, publishing headlines like:

- "Amazon Founder Says His Company 'Will Go Bankrupt'..."[37]
- "As Amazon CEO, Jeff Bezos was obsessed with the company's 'inevitable' death"[38]

At first glance, it looked like standard clickbait, but it was more than that. The press seemed genuinely shocked that Jeff had acknowledged Amazon would eventually fail. It showed how rarely companies talk about uncomfortable but obvious truths—like the fact that businesses don't last forever. Imagine if Jeff said, "I'm a human being, and someday I'm going to die." The headlines would have read, "Jeff Bezos Declares That His Death Is Certain!" Of course he's going to die someday, but there's this quiet, almost unconscious belief that some people and companies will be able to live forever.

This Leadership Principle is all about seeking the truth and seeing the world for what it is, even when it's uncomfortable.

It's Not "Be Right All the Time"

When I first read the Leadership Principle Are Right, A Lot, I thought it meant something like "Be right all the time." There's a big difference, though. No one can actually be right all the time. You can pretend to be right—like a parent who says, "I'm right because I said so." Being right a lot isn't about clinging to the illusion that you always have the answer—it's about knowing when you don't, being willing to admit it, and using that to get better.

---

[37] Jared George, "Amazon CEO Says His Company 'Will Go Bankrupt'...", *The Motley Fool*, accessed February 26, 2025, https://www.fool.ca/free-stock-report/amazon-ceo-says-his-company-will-go-bankrupt/.

[38] Isobel Asher Hamilton, "As Amazon CEO, Jeff Bezos was obsessed with the company's 'inevitable' death," *Business Insider*, December 6, 2018, https://www.businessinsider.com/jeff-bezos-keeps-talking-about-amazons-inevitable-death-2018-12.

People love the idea of some magical oracle that can predict the future and that companies like Amazon have special access to it. Management consultants, particularly firms like McKinsey, often claim to have even greater access to the oracle. I had a friend whose company hired McKinsey to build them a sales strategy. I have no problem with that. The issue was that seven years later, they were still using the same strategy. McKinsey created it, so it was treated like a sacred text—unchangeable, unquestionable, and somehow always right, even when reality said otherwise. They stuck to it like gospel, never mind that it was written by a few consultants barely old enough to rent a car.

Even as they watched the market shift, they clung to the illusion of certainty rather than adapt.

The human desire to predict the future goes back thousands of years. In ancient Greece, people traveled great distances to consult the Oracle at Delphi, hoping for divine answers. While no one knows what really happened there, here's one main theory: Petitioners would travel from all over Greece to get answers from Pythia, the priestess of the Temple of Apollo, who was believed to be a conduit to the gods. They would come with their urgent problems about the future from the state of the harvest to the outcome of a war. She would only listen nine days a year, once a month, never in winter. After hearing a request, she would enter the temple's inner sanctum, where a crevice in the ground released intoxicating vapors. There, after getting drunk on the fumes, she would channel the voice of Apollo, speaking in cryptic riddles and phrases. These messages were then interpreted by priests, who transformed them into the poetic hexameter of the day.

Petitioners left believing they had received a clear and definitive answer from the gods. In reality, the most famous predictor of the future was likely delivering incoherent, intoxicated ramblings—dressed up by priests to make it sound profound.

Being right a lot isn't about pretending to have all the answers. It's about staying curious, questioning assumptions, and adapting when the world changes. Those who believe they already know everything stop looking for the truth. And that's when they start getting things wrong.

Making Better Decisions

Are Right, A Lot means making better decisions over time. The goal is to do this in the most effective way—gathering the most valuable information at the lowest cost. When we have to navigate uncertainty and push into the unknown, we build a Minimum Viable Product (MVP), as discussed in **LP #2: Ownership**, to test our assumptions quickly and learn what works before making bigger commitments.[39]

But often there's an easier way to make a decision. Jeff says, "Many of the important decisions we make at Amazon.com can be made with data. There is a right answer or a wrong answer, a better answer or a worse answer, and math tells us which is which. These are our favorite kinds of decisions."[40]

Often the data already exists inside the company. I've found entire presentations with exactly what I needed. My favorite is when people say, "Here's my design for this application. We didn't have enough money to build it, but it's a start."

When I was at one company, I used an unconventional method to test a hypothesis—without building anything. A startup approached us with new technology that allowed clients to use their voice to log into their computers. The client would:

1.   Visit a website.
2.   Click a button to log in.
3.   Receive a phone call.
4.   Say the phrase "My voice is my password" to log in.

---

[39] Jeff Bezos, *Invent and Wander: The Collected Writings of Jeff Bezos* (Boston: Harvard Business Review Press, Kindle Edition, 2020), p. 72.

[40] Jeff Bezos, *Invent and Wander*, p. 72.

One of my fellow product managers was excited about this. "Our competitors are already piloting this technology," he said. "Our clients will think we're less innovative if we don't have it. Plus, even if we fail, it'll help build our innovation muscles."

We talked about why he wanted to do the pilot. He explained that he wanted to gather customer feedback on the technology. He wanted to do an MVP to see if this technology was a good fit for users. But after listening to him, I realized that the data he needed to make a decision already existed. Instead of building our own pilot and spending our own money, we could use the competitor's pilot for this! Many of our customers used both systems, and the technology was essentially the same.

So I started asking customers, "What do you think of this technology? Is it something we should build too?" Their responses were clear:

- "It's an interesting technology but it doesn't work a good portion of the time. That becomes very frustrating."
- "It's annoying to deal with a phone call when I'm already on my computer."
- "It feels awkward to say 'My voice is my password' while sitting at my desk in a cubicle."

So we were able to make our decision from the work done by our competitor. We were able to save hundreds of thousands of dollars and redirect our efforts toward solving more important customer problems.

Tell Me About a Time You Made a Mistake

Interviews at Amazon center on how candidates have applied the Leadership Principles in the past. For Are Right, A Lot, an interviewer might ask, "Tell me about a time you made a mistake." Naive candidates may freeze up, assuming they're being asked, "Give me a reason not to hire you." While this may be true at some companies, that's not the purpose of the question at Amazon. What the interviewer really wants to know is, "Do you understand that you're not perfect, and are you continually working to get better?"

The biggest mistake candidates make is trying to sidestep the question and act like they have nothing to improve. It's like the dreaded "Tell me about your greatest weakness" question. The worst answer? "I'm a perfectionist." It screams that you either don't know your own flaws or just refuse to admit them.

Imagine how that conversation might go:

> "Tell me about a weakness that you have," says the interviewer.
>
> "I'm a perfectionist," says the candidate.
>
> "How does that work as a weakness?" says the interviewer.
>
> "Well, I always spend a lot of time making sure everything is perfect," says the candidate.
>
> "That sounds difficult. Does that mean you don't get much done because you're always checking your work? Or does it mean you don't prioritize the most valuable tasks first? Or maybe you don't share interim work products with stakeholders. How does this weakness affect your work?" says the interviewer.
>
> "Ummm..." the candidate trails off.
>
> Don't pretend you're perfect. No one is perfect, and pretending otherwise shows that you don't know how to improve. Instead, be honest about a mistake you've made, explain what you learned from it, and demonstrate how it helped you grow.

An Even Worse Way to Answer This Question

Another version of the "What have you learned?" question is, "Tell me about something surprising you learned from a customer on a project." Here's an actual answer I received a few years ago.

He started off well. "We were working on a new trading system for a bank. Previously, traders had six different systems, each with its own window. We looked at what they were trying to accomplish, and we could put all of these systems together so that traders could have a much more holistic view of their data. However, when we showed it to the users, they didn't like it and wanted the six separate windows back."

"So what did you learn from that?" I asked.

"I really like talking to customers, and I'm always learning things from them. In this case, I learned that customers don't realize when a solution is better for them."

This was a big red flag. It was worse than saying he was a perfectionist because he understood the purpose of the question—he knew he was supposed to talk to customers and learn something—but he didn't take away anything useful from the experience. Instead, he concluded, "Sometimes other people don't know what's good for them." He could have said, "We needed to build something that still looked like the old system, at least as an option, because it takes time for people to change." If he dug deeper, he might have come to an answer like, "Traders have an enormous amount of information. Having these different applications allowed them to segment the different information into specific areas. Combining them all into one application removes this older, if haphazard, information design."[41]

The Right Way to Answer the Question

To answer this question effectively, you need to identify a development need and show how you're improving. Think of it as, "What are you working on?" You don't want to just say, "Here's a problem, and I need

---

[41] I learned about this when talking with a team looking at monitors for traders. Traders have four 32-inch monitors to give them the equivalent of a giant screen. When asked, "Why don't we just get you a giant screen?", they say, "I like the four screens better. The frames on the screens allow me to better segment my information."

you to fix it." Instead, you want to demonstrate, "I'm already working on it, and here's how."

If you do this well, your weakness starts to look like a strength in progress.

Here's one way I've answered this question in the past.

Example: One Time I Made a Mistake

I'm passionate about serving customers, and when I see an opportunity to make things better, my instinct is to move fast. But sometimes, that urgency turns into impatience—pushing forward before stepping back to prioritize. In my eagerness to help, I've chased solutions that mattered, but not always the ones that were the highest priority at the time.

Once, I was working on an analytics project that displayed the status of different initiatives. The goal was to redesign the interface so users could quickly understand which projects were on track and which ones needed help. As part of this, we focused on improving the layout of the projects, making it easier to identify patterns and spot issues at a glance.

When I reviewed the updated layout, I saw that it was a big improvement—but I also saw an opportunity to make it even better. I mentioned to the team that the icons were different sizes and it would make it prettier to standardize on a single size. I assumed this was a minor tweak, so I sent the developers on their way and waited for the change. A week later, I talked to the development team and asked what was taking so long.

"Rob," one of the engineers said, "we've finished most of the work, but we're running into spacing issues. The problem is that these icons are used in multiple places. When we resized them, it caused a ripple effect across the system. I've already fixed four of these issues—now I'm working on the fifth."

"Stop," I said. "Don't work on this anymore. It was just a suggestion. I didn't mean for you to prioritize this over other work."

I realized that, as a product manager and leader, there's no such thing as "just a suggestion" from the boss. What I had meant as a casual idea, the team took as a directive—shifting focus away from more critical tasks.

Now, I approach things differently. Instead of making suggestions during the review itself, I review completed work against the original requirements and add any additional improvements to the prioritization queue. Now I say, "I really like what you've done—let's get this out to customers. As a future improvement, let's look at changing the icon size." This small shift in wording would have allowed the team to prioritize their work better.

By answering in this way, I demonstrate self-awareness, an ability to recognize mistakes, and a commitment to improvement. That's the real purpose of the question: "Are you aware of your mistakes, and do you work to fix them?"

At the end of an interview, I like to turn this question around on my interviewer. I ask, "What's something surprising you discovered while building your product?" Customers rarely behave exactly as expected, so surprises happen all the time.

People who love learning usually light up when they hear this question. They can't wait to share what they've discovered, what challenged their assumptions, or what made them rethink their approach. Others, not so much. Either way, their response tells me a lot—whether they embrace curiosity and adaptability or prefer to stick to what they already know.

Final Thoughts

Amazon is obsessed with getting to the truth. To do that, it has to be comfortable with being wrong and not having all the answers upfront.

It thrives on curiosity, on constantly testing assumptions, on being willing to challenge its own thinking. This principle—Are Right, A Lot—isn't about always having the correct answer from the start; it's about developing the instincts and judgment to find the best answer over time.

The best decision-makers don't rely on gut feelings alone, nor do they cling to outdated models simply because they once worked. They seek out diverse perspectives, look for disconfirming evidence, and remain open to changing their minds when new information emerges. They know that the moment they stop questioning, stop learning, and start assuming they've got it all figured out, they've already started to lose.

Being right a lot isn't about confidence—it's about humility. That's why the question "Tell me about a time you made a mistake?" is so important. It's not a trap; it's a test of self-awareness. Strong decision-makers don't just get things right—they recognize when they've been wrong, learn from it, and adjust. The best leaders don't cling to bad ideas out of pride or fear of looking weak. They welcome challenges to their thinking, embrace feedback, and pivot when necessary.

The reality is, no company, no leader, no individual is right all the time. But those who consistently make better decisions do so not because they have some innate ability to predict the future, but because they embrace uncertainty, test their ideas, and learn from every unexpected result.

Or as Isaac Asimov is said to have remarked, "The most exciting phrase to hear in science, the one that heralds new discoveries, is not 'Eureka!' but 'That's funny...'"

LP #5

LEARN AND BE CURIOUS

Leaders are never done learning and always seek to improve themselves. They are curious about new possibilities and act to explore them.

"If you think of it in terms of the Gold Rush, then you'd be pretty depressed right now because the last nugget of gold would be gone. But the good thing is, with innovation, there isn't a last nugget. Every new thing creates two new questions and two new opportunities."

—Jeff, "The Electricity Metaphor," 1998 TED Talk

You might assume that everyone at Amazon would be naturally curious—after all, these are some of the smartest people in the world. But **LP #5: Learn and Be Curious** isn't about being smart; it's a mindset. It's about the desire to learn, especially in areas outside your comfort zone. It's about having a bias toward "tell me more" instead of "that's not my job."

I saw this during AWS onboarding. Everyone needed to learn the basics of how the cloud worked. This made sense for me as the Head of Banking—I had to understand and sell these products. But even new marketing hires from American Express or Prudential had to learn the fundamentals of AWS. They weren't cloud engineers, yet they all did it! It wasn't that they were smarter than anyone else; it was that they were curious and open to learning.

Curiosity in Action

My friend Rick was interviewing at Amazon. He was an investment banker—well over six feet tall, dark complexion, lanky, always dressed in fine Italian suits. He didn't look like a tech guy at all. But I knew he had the right mindset for Amazon. I wanted him to show off his love of learning and curiosity. So I told him to build a simple website on AWS and display a few lines of text.

He looked at me like I'd suggested we build a time machine in the garage. "Wait... what? Me?! I have no idea how to code!"

I explained that while it was intimidating, it wasn't that hard. All he had to do was create a basic "Hello World" program. A "Hello World" program is as simple as it gets. It's like turning the key in a car just to prove the engine works—nothing fancy, just a first step. There were plenty of online tutorials, and if he followed them step by step, he'd get there. He walked away skeptical but agreed to try.[42]

A week later, much to his own surprise, he sent me a link to a website that proudly displayed "Rick is Awesome!" in large font.

That's what curiosity is all about. Rick learned not to be intimidated by technical jargon like EC2 and S3.[43] By the time he was done, he could confidently say, "I created my own AWS EC2 instance, then set up a public S3 bucket to host my website." Suddenly, this banker with zero technical experience sounded like a seasoned programmer.

---

[42] My favorite AWS tutorial was the AWS Cloud Practitioner certification from the training site acloud.guru.

[43] EC2 is a fancy name for a computer in the cloud. S3 is a fancy name for a piece of data that you store in the cloud. In an ironic naming twist, S3 technically stands for Simple Storage Service, but let's be honest—what started as a clear, approachable name has now become just another piece of tech jargon.

Learning Should Be Fun

To learn something new, you have to accept that you don't know everything—otherwise, there's nothing to learn. But that's easier said than done. Imagine you've spent months perfecting a presentation, refining every detail, only to get an email from your boss on the morning of the big meeting: "I'm out sick. You know this material better than anyone. You've got this."

Your stomach drops. Your hands get clammy. This isn't how today was supposed to go. You wanted your boss to present it, to handle the tough questions. Now, instead of feeling proud of your work, you just want to call in sick too.

But you could process this differently. Fear and excitement feel exactly the same in the body—it's all about how you frame it.[44] Imagine two brothers at the top of a five-story water slide. The first one looks down and panics. "This is so high. My heart is racing. Nope. No way. Not doing it." The second brother feels the exact same adrenaline rush but grins. "This is so high. My heart is racing. This is going to be amazing. Let's go!" Same slide, same feeling—completely different mindset.

Some Leadership Principles demand a narrow, focused approach—like **LP #7: Insist on the Highest Standards** and **LP #14: Deliver Results**. But when it comes to this principle, you need to zoom out, stay open, and embrace the ride.

Being in the Moment

Learning in a relaxed environment can be enormously fun, but if you put too much pressure on yourself, even fun things can become frustrating. Take my friend Dan. Dan is a senior leader who rose to

---

[44] According to the Yale Center for Emotional Intelligence, emotions have two components: pleasantness (how much you like it) and energy (how actively engaged you are). You can learn more about this at the Yale Center for Emotional Intelligence's website (https://www.ycei.org) or by exploring their research on the Mood Meter, a tool designed to help people recognize and regulate emotions more effectively. Their work emphasizes how understanding these two dimensions of emotion can improve decision-making, relationships, and overall well-being.

the highest levels of his organization before he was 40. He worked hard, and it paid off.

He was telling me about his upcoming vacation to Florida.

"We'll probably go to the beach and throw a frisbee around. It's so frustrating, though, because my six-year-old doesn't throw the frisbee well," he said. "I keep showing him what to do, but he never gets better."

I told him to take a different perspective. Instead of focusing on the outcome, I suggested he enjoy the process of watching his son grow and learn. It's fun and rewarding to see kids get better. Soon his son would be throwing better than him.

Dan had turned something that should be fun into work. Lots of people do this. Timothy Gallwey wrote the bestselling book *The Inner Game of Tennis* about this very problem.[45] While coaching people on weekends, he noticed that tennis, something meant to be enjoyable, had become an obsession. Players cared only about winning and berated themselves over "stupid" mistakes. They sucked all the joy out of their weekend leisure activity. Worse, by beating themselves up, they made it harder to improve.

Gallwey argued that the trick to both enjoying the game and improving is to keep an open mind and stay in the moment. He described a state of peak performance—what others call "flow"—where you don't consciously think about how to hit the ball; you just trust yourself to do it.

As a teenager, I played ping pong against my cousin Richard. Being competitive, I wanted to win. I didn't care whether I won the point because I played well or because he made a mistake. Either one worked, as long as I won. My enjoyment of the game depended entirely on the outcome. Only one of us could be happy at the end of the match.

It was like when little kids finish a soccer game and are forced to shake hands and say, "Good game," when only the winning team really means it.

<hr />

[45] Timothy Gallwey, *The Inner Game of Tennis: The Classic Guide to the Mental Side of Peak Performance* (New York: Random House, 1997).

But over time, I realized I needed to focus on playing rather than whether I was winning. As I got older, I paid more attention to the feeling of hitting the ball. Sometimes it went in, sometimes it didn't. But by focusing on the moment—on the ball itself and the rhythm of the game—I enjoyed it so much more and got much better.

I bring the same mindset to my work, approaching each task with a sense of playfulness. My kids like to tease me about it—"Why is it that whenever we ask what you did today, you always talk about how much fun you had?"—but it's true. Even in frustrating situations, I look for a way to reframe them. If I'm struggling to track down a hard-to-find person or a key piece of data, I turn it into a kind of treasure hunt, an adventure where the missing detail is the final clue. If I'm navigating an organizational challenge, I picture a chessboard, moving pieces strategically to figure out how we can win. Shifting the perspective doesn't change the work itself, but it makes the process more engaging—and often, more effective.[46]

How to Master the Seemingly Impossible

When I was at Yale, I was surrounded by incredible people doing amazing things. There were artists and writers putting on shows and publishing books. Math geniuses. Students building solar-powered cars in the middle of the night. I could never do any of those things.

And then there were the jugglers.

Every Sunday on Old Campus, a group of students gathered to toss around balls and clubs in impossibly complex patterns. The thing that stood out? None of them had arrived at college as jugglers. They weren't athletes, either. Many of them were in the band. Though they didn't have the hand-eye coordination to throw a perfect fastball (or any sort of fastball), they had no problem passing bowling pins and knives to each other.

[46] The book *Level Up Your Life: How to Unlock Adventure and Happiness by Becoming the Hero of Your Own Story* takes this to a new level, helping you rethink planning your life as a game of Dungeons and Dragons.

That inspired me. If these amateurs could master such a skill, why couldn't I?

So I picked up a book called *Juggling for the Complete Klutz*[47] and started practicing. Clearly, based on the title, the authors believed anyone could learn how to juggle.

It's surprisingly simple if you put in the effort. I spent three weeks tossing balls in the air in my dorm room, and before long, I could juggle three balls pretty well. I was so proud of myself that I told my friends.

They all went through the same transformation:

1. That looks really hard. I could never do it.
2. But Rob just learned... maybe I should try.
3. It's a bit of work, but I'm making progress.
4. Wow, I can juggle! I want to show everyone!

It's not about talent or innate ability. It's about mindset. Focus on the process, not just the outcome. And most of all, have fun with it.

I Wonder If...

Curiosity starts with a question. It's about asking, "I wonder if..." or "I wonder how..." and then looking for the answer.

When I was in college, I would set off on little quests with my friends Lutz and Christine, chasing down odd questions just to see where they led. "Is there a castle on top of the library?" It turns out there is—built to disguise the rooftop machinery.[48] "What's the most interesting book they'll actually let us hold in the rare books library?"

---

[47] John Cassidy and B. C. Rimbeaux, *Juggling for the Complete Klutz* (New York: Klutz Press, 1977). This book started as a pamphlet to teach school kids to juggle and is the first book of the Klutz publishing empire. This pre-dated the "For Dummies" and "Complete Idiot's Guide" series by decades.

[48] If you look at the top of Yale's Sterling Memorial Library, you can see something that looks like a castle (zoom in on the roof). Kids today can just go to Facebook and see much better pictures, posted by the university.

That one led us to something incredible: an early printing of *Hamlet*, complete with an inscription from Shakespeare himself. One of the only known samples of his handwriting—right there, in our hands.

One of my favorite "I wonder ifs..." is about a T-shirt.

In the 1990s, I read about a T-shirt that was classified as a weapon. At the time, the U.S. government was highly protective of "strong encryption"—the technology that allows people to send secure messages. It's how spies communicate without governments intercepting their messages and how people can safely transmit credit card information over the internet.

As late as 1992, cryptography was on the U.S. Munitions List as Auxiliary Military Equipment.[49] This meant that giving encryption code to a foreign country was legally equivalent to supplying them with weapons. But as the internet took off, this law became absurd. Secure online transactions require strong encryption.

By the mid-1990s, a group of computer programmers started playing with these restrictions, testing how small they could make an encryption algorithm. Eventually, they reduced it to just a few lines of code. Then someone had an idea: they printed the code on a T-shirt. According to the law, wearing the T-shirt on an overseas flight or even showing it to a foreign national was considered exporting a weapon.[50]

The law changed in 2000, and the T-shirt became a lot less interesting for people. It went out of print. So I asked myself, "I wonder if I could make this T-shirt myself?"

I found a wonderful design by Vipul Ved Prakash that arranged the code in the shape of a dolphin. ThinkGeek printed the shirt in the 1990s, but since it was no longer available, I decided to print one myself.

Once I started building things, it felt like I had a superpower. I could learn things I once thought were impossible. I could create things I wanted and bring them to life. Instead of saying, "This is impossible," I started asking, "I wonder if I could do this?"

---

[49] "United States Munitions List," Wikipedia, https://en.wikipedia.org/wiki/United_States_Munitions_List.

[50] See the full details on my website at https://schlaff.com/wp/how-i-re-built-my-favorite-t-shirt/.

Final Thoughts

The people who thrive at Amazon are curious. No matter where they come from—banking, marketing, engineering, or beyond—they're there because they want to build something new, to push boundaries, to learn. Just like Rick, the investment banker who built a website on AWS, they don't let a lack of experience stop them. They just dive in.

But curiosity isn't about doing something just because a company or a boss says you should. It's about finding the fun in it, turning learning into an adventure instead of a chore. The best learning happens when you stop worrying about whether you're getting it right and start enjoying the journey.

To me, this allows me to ask the question "I wonder if..." with a sense of possibility instead of doubt. "I wonder if I could learn how to juggle." "I wonder if I could recreate my favorite T-shirt." Or, "I wonder if I could do this thing that seems impossible."

I'll end with a skill that's incredibly powerful but frightening to many people: drawing. I'm not saying that everyone needs to be the next Picasso or Monet. But having the ability to make simple drawings on a whiteboard at work can change the way people think. Here's a sample of how well (or poorly) I draw. I'm trying to communicate how pictures can lead to powerful ideas.

An example of my drawing—pictures create powerful ideas

I had always been uncomfortable with my drawing abilities. Then I had a conversation with Katie, my wife's 12-year-old cousin. Even at 12, Katie was quite the artist. Her grandmother, Margo, is an acclaimed painter who sells in local galleries and teaches art classes in the south of France.

Over dinner one night, I told Katie about my drawing fears. She said, "Oh, if you want to draw, you should check out Ed Emberley. His books can help you out." These books were exactly what I needed. They used simple shapes to create faces and trains and everything under the sun. He even has a book called *How to Draw a World*.

As I kept practicing, people would say, "You draw well. I could never do that!" But the truth is, I don't draw that well—I just draw. And that's the difference. Anyone can do it once they get past the fear of trying. You know how I know this? Remember that book by Ed Emberley that Katie recommended? It's a children's book. For ages five to eight.

LP #6

HIRE AND DEVELOP THE BEST

Leaders raise the performance bar with every hire and promotion. They recognize exceptional talent and willingly move them throughout the organization. Leaders develop leaders and take seriously their role in coaching others. We work on behalf of our people to invent mechanisms for development like Career Choice.

I'd rather interview 50 people and not hire anyone than hire the wrong person.

—Jeff, "Inside the Mind of Jeff Bezos," Fast Company, August 1, 2004

A candidate steps on-site for their day of interviews—"The Loop." An interviewer takes out their laptop and starts taking meticulous notes. This is so awkward that most interviewers feel the need to apologize, saying something like, "Before we begin, I'd like to apologize for taking notes as we talk. I assure you that I'm listening, but at Amazon, we take copious notes so that we don't miss anything."

For a company that prides itself on efficiency, Amazon has a surprisingly heavyweight hiring process. At most other companies I've worked at, hiring is one of the least structured processes—managers often have the freedom to hire as they see fit. But at Amazon, this process allows for faster and higher-quality hiring.

Why go through all of that? Don't they have a million great candidates lining up to work there? Can't it just skim off the cream of the crop? You'd think so, but things don't work that way. Amazon isn't great because it magically attracts top talent. It's great because it hires the right people. This chapter will show you how.

You might be wondering, "Do interviewees even want to put up with this?" While it may seem daunting on paper, think about how a standard hiring process works. A friend of mine interviewed at a digital marketing company. She told me the process was grueling—five rounds of interviews, with at least two people in each. She said, "It's not like this company was Amazon, where everyone wants to work. They should be nicer to their interviewees."

The irony is that Amazon is nicer to candidates than that digital marketing company. In fact, Amazon is often nicer to its interviewees than to its employees. It strives to give each candidate a great experience, whether or not they get the job. It knows that virtually everyone who interviews is also an Amazon customer, and it doesn't want to lose a customer over a bad interview experience.

Why Not Leave It up to the Hiring Manager?

Why does Amazon go through all this effort? Why not just let hiring managers decide who to hire? If hiring is the most important decision at the company, shouldn't we remove as much process as possible to get the best people?

Yes, hiring is critical. But it's also an emotional decision, which is exactly why it needs a strong process—to remove emotion from the equation.

A hiring manager has to balance many competing needs. The candidate needs to be good at the job. They need to fit with the team. They need to align with the company culture. And ideally, they need to start quickly because the hiring manager is already overworked.

Trying to juggle all of these factors in real-time is difficult. We like to think we can take everything into account and just "feel" who the best candidate is, but as it turns out, emotionally charged, complex decisions are the worst ones to trust your intuition on.

Daniel Kahneman, the Nobel Prize-winning economist, writes about this in *Thinking, Fast and Slow*. He explains that our brains have two decision-making modes: an intuitive, gut-feel mode and a thoughtful, reflective mode. A structured hiring process forces our brains into the reflective mode, helping us avoid emotional pitfalls and biases. Relying on a strong process allows hiring managers to focus on the candidate and make better decisions.

The Bar Raiser

In the late 1990s, Amazon had a problem. The company was growing fast—really fast. It expanded from 600 employees in 1997 to 9,000 by 2000. In the early days, Jeff interviewed every candidate himself. But as the company scaled, this became impossible. By 2000, only a tiny fraction of employees were interviewed by Jeff.

At first, Amazon hired the way most companies do—by feel. The assumption was that if someone was good enough to be hired by Amazon, they would also be good enough to hire others.

This wasn't working.

In the late '90s, Amazon brought in a new director from a larger tech company. He began staffing his team with people he had previously worked with—people who got a bit of a free pass because he knew them. Over time, he built an entire team of subpar employees. It became clear that if Amazon was going to maintain its quality while scaling, it needed a systematic way to replicate Jeff's original hands-on interview approach.

So Amazon created a mechanism to ensure it was hiring the right people. In most companies, the hiring manager has the final say, with HR and recruiting acting as support. But Amazon wanted someone to actively fight for the company during the hiring process—someone who could represent Amazon's long-term interests with equal or even greater weight than the hiring manager. That's why it created the Bar Raiser.

The Bar Raiser is a senior interviewer who represents Amazon and its Leadership Principles in the hiring process. They interview candidates alongside the rest of the team and run the debrief meeting where hiring decisions are made. They even have the power to veto a hiring decision if necessary. While this veto power is rarely used, its existence ensures that hiring managers stay accountable.

Amazon actually borrowed this concept from Microsoft, where seasoned interviewers unaffiliated with the hiring team—known as "AA" (short for "as appropriate")—acted as the final interviewers. Their role was to ensure hiring standards were upheld across the organization. However, at Microsoft, the AA were just a part of the hiring process. At Amazon, the Bar Raiser became its core.

Breaking Down the Process

So what does Amazon's hiring process look like today? It consists of the following steps:

- Job Description
- Assessments and Resume Review
- Phone Screen
- The Loop
- Written Feedback
- Debrief/Hiring Meeting
- Communicating the Decision

I won't go into the details of each step, but I do want to highlight how this process works. It's designed to be precise, thoughtful, and reliable—ensuring hiring decisions are based on more than just gut instinct.

These steps fall into three key categories:

1. Define the bar for the role
2. Measure candidates against that bar
3. Make a decision and communicate it

Define the Bar for the Role

You can't get what you want if you don't know what you're looking for. When creating a job description, you're setting the bar to measure candidates against.

In most companies I've worked for, job descriptions contain basic facts and serve as a way to cast a wide net for candidates. The hiring manager assumes that once they meet people, the process will take care of itself. But writing a clear and concise job description establishes a solid benchmark for evaluating candidates. Without that benchmark, hiring managers end up relying on gut instinct or resorting to a vague approach of "getting the best people we can." The more we define up front, the better we can assess and compare candidates in the next phase.

Measure Candidates Against That Bar

The next stages of the hiring process involve increasingly rigorous evaluations. It starts with a resume review and possibly some online assessments. At this point, Amazon is looking for candidates who, on paper, have the necessary skills and a track record of delivering results.

Then comes the phone screen. A hiring manager or designated interviewer conducts a one-hour call with the candidate to assess their fit for the role and alignment with Amazon's Leadership Principles.

This step helps filter out candidates early, ensuring that only those who show promise move forward. It allows Amazon to cast a wide net without wasting valuable interviewing resources.

Next is "The Loop"—a one-day gauntlet of interviews designed to answer two key questions:

1. Can the candidate do the job?
2. Do they fit with Amazon's culture?

For both questions, interviewers use behavioral-style questions: "Tell me about a time when…" to understand how candidates have handled real-world situations.

When it comes to job capabilities, candidates are measured against the bar set by the job description. Do they have the required skills? But at Amazon, just meeting the bar isn't enough. Candidates need to be better than 50% of the people already in that role.

Cultural fit is just as critical. Each interviewer evaluates the candidate against a couple of Amazon's Leadership Principles, using behavioral questions to gauge alignment. This is where the feverish note-taking comes in. Interviewers document specific examples from the candidate's career to determine how well they match Amazon's principles.

Making and Communicating a Decision

After The Loop, interviewers enter their evaluations. But there's a key rule: interviewers must submit their feedback in writing before the meeting. They can't discuss the interview or see each other's reviews until their own is finalized. This ensures independent assessments, preventing groupthink. Once feedback is submitted, interviewers can view other evaluations in an online portal—seeing which colleagues were "inclined" (in favor of hiring) or "not inclined" (against hiring), along with their reasoning.

At the debrief meeting, everyone gives their rationale for whether the candidate meets the bar. The Bar Raiser leads the discussion, ensuring each interviewer presents their assessment before the group deliberates. This is where the process really shines—decisions aren't made on a whim or dragged out indefinitely. By relying on data and a structured evaluation, Amazon moves fast. Within five days of The Loop, candidates receive a final decision.

If they meet the job requirements, align with Amazon's culture, and outperform at least half of the people already in the role, they're in. It doesn't matter if someone even better is interviewing next week. If this candidate meets the bar, they get the job.

Speed matters. I saw this firsthand at a financial services company. Our customer quality was lower than our competitors', and the reason was simple—we were too slow. We wanted to get our credit and pricing decisions perfect, and it took us forever to make a final proposal. While we spent time fine-tuning, the customers were moving on. The ones who chose us were the ones who had no better options.

By overanalyzing every decision and aiming for perfection, we didn't get better results—we got worse ones. Amazon avoids this trap. It doesn't chase "perfect." It focuses on fast, high-quality hiring decisions, backed by a clear and structured process.

Benefits of Amazon's Hiring Process

Amazon's hiring process leads to better decision-making by avoiding many of the pitfalls common in traditional hiring.

Not Hiring Just to Fill a Spot

Put yourself in the hiring manager's shoes. You work for a company that wants to keep costs low and ensure everyone is working at maximum capacity—otherwise, it's a waste of money. You're already stretched thin, covering your own responsibilities while also handling the work of the vacant role. You need to hire someone fast to lighten your load.

The temptation is to rush the process and hire the first candidate who seems good enough. But hiring too quickly increases the risk of bringing in the wrong person, which ends up costing the company far more in the long run—not just in salary, but in lost productivity, additional training, and potentially having to start the hiring process all over again if they don't work out.

Not Hiring Based on "Gut Instinct"

Many companies leave hiring entirely up to the hiring manager, assuming they know best. It seems logical—after all, they understand their team's needs. But without a structured process, hiring managers often fall back on their gut instincts, which invites bias into the process.

Your gut can lead you astray in countless ways. I remember interviewing a candidate for another hiring manager. I followed a structured evaluation process and found that the candidate performed poorly—he lacked a strong grasp of product management and wasn't effective at managing stakeholders, the two most critical aspects of the role.

But the hiring manager was enamored. "But he went to MIT!" they said. "Wouldn't it be great to have someone on the team who went to MIT?!"

Rather than focusing on whether the candidate could actually do the job, the hiring manager was dazzled by a prestigious credential. This is exactly why a structured process matters—it prevents us from making decisions based on arbitrary factors rather than whether they can do the job.

Not Waiting for the Mythical "Best" Person

Once hiring managers start interviewing candidates, they face a difficult decision-making problem. Without a defined process, they tend to compare each new candidate to the ones they've already seen rather than to an objective standard.

This often leads to indecision. There's always the possibility that someone better will come along next week, so hiring managers hesitate, dragging the process out indefinitely. As a result, strong candidates are left waiting, or worse, they drop out entirely because the process feels long, unclear, and frustrating.

Final Thoughts

Hiring is one of the most important decisions a company makes, yet most companies treat it as an afterthought. I've seen hiring managers chase the "best" candidates without ever defining what best actually means. It's a vague, feel-good goal that often leads to hiring people who look great on paper but don't necessarily fit the job.

Amazon takes a different approach. Instead of endlessly searching for some theoretical *best*, it defines the role clearly, sets a measurable hiring bar, and selects candidates who meet that bar *and* outperform at least half the people already in the role. It's not about chasing superstars— it's about making the team stronger with every hire.

I learned that hiring the right people for a role is more important than simply hiring the "best" in a general sense. At one company, I worked with a manager who led a 15-person customer support team. Her boss insisted she only hire "A players," but she pushed back.

"These are customer support roles," she told me. "I need people who do their job *exceptionally* well from 9 to 5. There's one promotable job on this team—mine. If we fill these roles with ambitious people who want to climb the ladder quickly, they'll just get frustrated and leave. I'd rather hire people who love the job they were hired to do."

That kind of clarity was rare. A friend once joked that our company had "the highest intellect per unit of output," which sounded flattering until you realized it was a backhanded compliment. We had plenty of brilliant people, but we couldn't get anything meaningful done. Amazon gets this: hire for the role, not the resume. The best hire isn't always the flashiest—it's the one who actually gets the job done.

LP #7

INSIST ON THE HIGHEST STANDARDS

Leaders have relentlessly high standards—many people may think these standards are unreasonably high. Leaders are continually raising the bar and drive their teams to deliver high quality products, services, and processes. Leaders ensure that defects do not get sent down the line and that problems are fixed so they stay fixed.

The football coach doesn't need to be able to throw, and a film director doesn't need to be able to act. But they both do need to recognize high standards for those things and teach realistic expectations on scope.

—Jeff, Amazon Shareholder Meeting, 2018

When I first saw this Leadership Principle, I wondered how anyone could succeed at Amazon. Insist on the Highest Standards made it seem like nothing would ever be good enough. Amazon's standards are high. As the LP states, many people think they're unreasonably high. But it isn't about making things perfect—it's about trying your best, taking pride in your work, and fixing things immediately when they break. It's about continually improving the quality of everything at Amazon.

This is why people trust Amazon Web Services (AWS) to run their IT infrastructure. If someone else is handling your IT, their failure is *your* failure. No one puts their business in the hands of a company that doesn't hold itself to the highest standards.

When I was at Amazon, I learned about a critical infrastructure service the company had built in-house. Curious, I asked a veteran Amazonian why they hadn't just used an industry partner. Was this about market share? Was Amazon trying to dominate another space?

"No," the engineer said. "There's really only one player in this game, and we tried partnering with them. But one Friday afternoon, their service failed. They told us they'd fix it on Monday morning. We told them that was unacceptable because the Amazon team can't go home until it's up and running. There was no one in the market who could provide it at the level of reliability our customers needed, so we had to build it ourselves." That's what Insist on the Highest Standards is about—never accepting "good enough" when customers are counting on you to get it right.

The Oxford Comma

During orientation, we were told about a new employee who was presenting in front of Andy Jassy, then CEO of AWS, for the first time. He was nervous. He'd spent countless hours writing his six-page narrative because he knew of Andy's commitment to high standards. Andy started reading the document, paused, stopped, and threw the young man out of the room.

What did the employee do to piss Andy off? Did he ramble on for more than six pages? Did he forget to spell-check? No—the young man forgot to use an Oxford comma.

You can be forgiven for not knowing what an Oxford comma is—or, if you do, for not caring much about it. The band Vampire Weekend even has a song that starts with the line, "Who gives a f*ck about an

Oxford comma?" But as grammar nerds know, an Oxford comma is the comma before the conjunction in a list of three or more parallel elements.

Without an Oxford comma, the sentence: "My heroes are my parents, Superman and Wonder Woman" could imply that your parents are superheroes, whereas: "My heroes are my parents, Superman, and Wonder Woman" makes it clear that they aren't.

"My heroes are my parents, Superman and Wonder Woman"

Everyone at Amazon knew about the Oxford comma. It was drilled into every new hire as part of their writing education—right up there with learning how to craft a six-pager. According to Amazon's doctrine, the Oxford comma wasn't just a preference; it was the right way to write. Clearer, more precise, and impossible to misinterpret.

An Oxford comma has its benefits, but in reality, it's a grammatical point of dogma. Some people love it, believing it enhances clarity, while others think it creates clunky, unnecessary pauses. The relative merits are fairly insignificant. The *Chicago Manual of Style* requires it. The *New York Times Style Guide* does not. Whether you use it or not, the important thing is consistency across the organization. But like any good doctrine, its adherents will fight to the death to defend it.

My Discovery

After being drilled at orientation about the Oxford comma and its importance, I—along with everyone else—used it without fail. That's why I was astonished to find the Oxford comma missing from Amazon's Leadership Principles.

It was missing from the one place it absolutely should have been. This is what the LP looked like in 2018:

Insist on the Highest Standards

Leaders have relentlessly high standards - many people may think these standards are unreasonably high. Leaders are continually raising the bar and drive their teams to deliver high quality products, services and processes. Leaders ensure that defects do not get sent down the line and that problems are fixed so they stay fixed.

—Amazon's Leadership Principles (circa 2018)

You can see (or rather *not* see) the missing comma after the word "services." And it wasn't just the Oxford comma. Once I noticed this, I found other punctuation errors in the Leadership Principles, such as using hyphens ("-") instead of em dashes ("—") to indicate pauses.

Now that I had seen these errors, the question was: "What do I do about it?" At most companies, it wouldn't even be a question. I'd show it to my friends, we'd have a good laugh, and no one would do anything about it. It would be like the case of the P.S. 199 school in Manhattan, which the New York City Department of Education calls "The P.S. 199 Jessie Isador Straus"—even though the school is named after the man Jesse Isidor Straus. The city managed to add an extra "i" to his first name and replaced the "a" with a "i" in his middle name. No one can figure out how to fix it.

At first, I thought, "These Leadership Principles are sacrosanct! They should not be touched!" (In my head, I heard those words in the cartoonish voice of God from *The Simpsons*.) But I wanted to test Amazon. The company claimed to embrace feedback and constant improvement—but did that extend to its own Leadership Principles, or was that just for everyone else?

So, who should I tell? The first suggestion I received was: "You could email Jeff. He is the ultimate owner of the Leadership Principles." While true, that felt like overkill. Surely there was an easier way.

While researching the LPs on Amazon's intranet, I found an internal wiki tracking edits. One Amazonian had made countless revisions, so I emailed him. Turns out, he was the guy—the official LP maintainer, an Amazon veteran of two decades, and someone who had been in the room during the last revision. At first, he was nervous, explaining that the senior team had carefully crafted the wording. But when I told him it was just a punctuation fix, he was relieved—and happy to update it.

Now, if you check the copy of the LPs on amazon.jobs, you'll see that the Oxford comma has been added! Older versions, still archived on archive.org, don't have it. Like every other company, Amazon makes mistakes. The difference is that it doesn't ignore them and works quickly to fix them.

How Did Everyone Miss This?

With all of Amazon's training, the missing Oxford comma should have been glaringly obvious. It should have flashed like a giant neon sign saying, *OXFORD COMMA MISSING!* But somehow, over 10 years, millions of people didn't notice.

When I pointed out the errors to other Amazonians, their first reaction was denial. "This couldn't be," they said. "Jeff and his senior leaders spent enormous amounts of time on this. They couldn't have missed it." But then they looked again—and started laughing, saying, "Wow. I can't believe I didn't see that before."

It's common for things to hide in plain sight. Scientists call it selective attention. One famous example is the "Invisible Gorilla Experiment"[51] (if you haven't seen it, check out the link). There's so much happening around us that our brains prioritize what's important—and simply fill in the blanks for the rest.

One of my favorite examples of this involves Cary Grant, one of the greatest film stars of Hollywood's Golden Age. Grant was an avid fan of magic and a founding member of *The Magic Castle*, a private club for magicians in Los Angeles.

When guests pulled up to The Castle, a doorman who looked shockingly like Cary Grant would step forward to open their doors. Everything about him—the face, the mannerisms—was identical to Grant. Visitors assumed it was an illusion, an elaborate trick of Hollywood makeup and acting.

But there was no trick. It *was* Cary Grant.

Grant wasn't part of a performance—he simply enjoyed chatting with the receptionist, Joan Lawton, who was studying child psychology. As a devoted father, Grant was fascinated by the subject. And when a guest pulled up, ever the gentleman, he'd instinctively hold the door open.

Even though guests saw him with their own eyes, the context threw them off. Their brains refused to accept that a movie star would be working as a doorman, so they did the only thing that made sense—they assumed he was just a lookalike.

[51] Christopher Chabris and Daniel Simons, *The Invisible Gorilla*, accessed March 8, 2025, https://www.theinvisiblegorilla.com/IGvideos.html.

Things hide in plain sight all the time. Take the Amazon logo. Have you ever noticed the arrow pointing from "A" to "Z"? Or the fact that it forms a smile? Once you see it, you can't unsee it.

So how did I see the missing Oxford comma? The answer: a shift in context. When I joined Amazon, I studied the Leadership Principles every day. I had an app on my phone to help me review them, but there was a problem—I couldn't simply copy and paste the text into the app.

So, I had to retype every principle manually, switching between the app and the website. Instead of just reading the LPs, I was forced to examine every single word—including the punctuation. And that's when I saw it.

Most people skim text, absorbing meaning without noticing the mechanics behind the words. We assume things are correct, especially when they've been scrutinized by senior leadership at a trillion-dollar company. But assumptions blind us. My experience retyping the LPs forced me to see what had been hidden in plain sight for years.

Trust but Verify

So why did Andy throw the guy out of the room for not using the Oxford comma? Did it really make the document that much worse? It wasn't that he cared deeply about the Oxford comma itself but rather what it represented. Amazon's standard is based on the *Chicago Manual of Style*, which mandates the Oxford comma. In a company with hundreds of thousands of releases a year, ensuring that every detail is checked is critical. Andy could use the simple heuristic of "Did you use the Oxford comma?" as a quick measure of whether someone paid attention to all the details.

David Lee Roth tells a similar story about Van Halen in the 1970s, when he was the band's lead singer. At the time, Van Halen was one of the first bands to tour across America, not just in major cities like Los Angeles and Chicago but also in smaller, third-tier markets, like Cincinnati and Cleveland, that had never hosted large-scale rock concerts. These productions required meticulous attention to detail—Van Halen's setup involved nine 18-wheelers carrying complex stage equipment. If the venue staff failed to follow instructions precisely, girders could collapse, or the floor could cave in.

Munchies

    Potato chips with assorted dips
    Nuts
    Pretzels
➡ M & M's (WARNING: ABSOLUTELY NO BROWN ONES)
    Twelve (12) Reese's peanut butter cups
    Twelve (12) assorted Dannon yogurt (on ice)

An extract of Van Halen's contract via The Smoking Gun[52]

To ensure that venues met their high standards, the band created a simple test. Buried deep within the 53-page rider that contained the band's requirements was an odd clause: they required a bowl of M&M's backstage, but *absolutely no brown ones*. If they walked into a venue and saw brown M&M's, it was an immediate red flag that the contract hadn't been thoroughly read. That meant they had to check *everything* to avoid potentially life-threatening mistakes.

This principle—trust but verify—is a common pattern for maintaining quality. In high school, I had a chemistry teacher who

---

[52] "Van Halen's Legendary M&Ms Rider," The Smoking Gun, December 11, 2008, https://www.thesmokinggun.com/documents/crime/van-halens-legendary-mms-rider.

assigned weekly homework but didn't check it regularly. Instead, he told us that once a semester, he would randomly check one assignment, and that one would count for 10% of our grade. Not wanting to risk it, we did every homework assignment.

Security Demands the Highest Standards

Think of computer security like living in a fortress. You can have steel doors, high walls, and guards at every entrance—but if someone leaves the outside door to the game room unlocked, none of it matters. Hackers don't need to knock down the front gate; they just need to find that one weak spot no one thought to secure.

And those weak spots tend to show up in the strangest places. In 2017, the security company Darktrace revealed how hackers broke into a casino—not through the betting systems, not through a high-stakes heist, but through an internet-connected fish tank.[53] The casino had installed a high-tech aquarium that could be monitored online, allowing staff to adjust temperature and feeding schedules remotely. The manufacturers focused on the smart features but didn't take security seriously. "Who would hack a fish tank?" they thought.

Hackers, that's who.

Once inside, they used the fish tank's internet connection to slip into the casino's broader network. From there, they tried to go after the real prizes—like the casino's list of VIP gamblers. It was a perfect example of how one small oversight can unravel an entire security system.

In the words of Steve Jobs, hackers "Think Different." But these hackers are out to cause some real trouble. The best way to stay safe? Think differently, too. That's why more schools are teaching these skills—not just to stop the bad guys, but to help students learn to approach problems from fresh, unexpected angles.

---

[53] National Cybersecurity Center, https://cyber-center.org/a-cyber-fish-tale/.

Teaching the Hacker's Mindset

I took Yale's most famous computer science course, CS223, with Professor Stanley Eisenstat. He taught us about how even when you think you've got everything tied up perfectly, a clever adversary can find ways to break in.

To illustrate, he played a game of Hangman with the class. He chose a four-letter word and gave us eight chances to guess it. To make it easy, he even provided the last three letters upfront:

_ill

We had eight guesses to figure out one letter. This should have been simple. The class enthusiastically started guessing:

"Bill!"

"Dill!"

"Fill!"

"Gill!"

"Hill!"

"Jill!"

"Kill!"

"Mill!"

And then we lost. So we tried again with different words:

"Pill!"

"Sill!"

"Till!"

"Will!"

...

And we lost again.

Game after game, the same result. At first, we chalked it up to bad luck. But then Professor Eisenstat revealed the real title of the class: "How to Cheat at Hangman."

There are 12 words in the English language that fit the pattern "_ill": bill, dill, fill, gill, hill, jill, kill, mill, pill, sill, till, and will. We assumed that he was playing with the traditional rules—where a single word had been chosen at the start of the game. But here's the trick—Professor Eisenstat hadn't picked just *one* word. Each time we guessed, he simply said "No" and mentally eliminated that word from the list. Once we had exhausted our guesses, he revealed one of the remaining words. No matter what we did, we were destined to lose. Clever adversaries don't play fair.

Some professors take this lesson a step further. At Harvard, one cybersecurity class assigns students an impossible task: memorizing the first 100 digits of pi in two days. The real goal isn't for them to memorize—it's for them to figure out how to cheat convincingly. The assignment forces students to think like adversaries, identifying every possible loophole because, in the real world, that's exactly what attackers do.

Students develop wildly creative solutions. One used invisible ink, readable only with special glasses. Another wrote the digits in Chinese, guessing the professor wouldn't be able to read them (he couldn't). The most ingenious solution came from a student who programmed his phone to transmit the digits in Morse code through vibrations. He sat in class, appearing deep in thought, as his phone silently fed him the answers.[54]

My Takeaways

Insist on the Highest Standards isn't about perfection for perfection's sake. It's about trust—knowing you can rely on others to do high-quality work. Customers trust Amazon to deliver Prime packages in two days. Banks and 911 centers trust AWS to stay online. That trust doesn't happen by accident.

---

[54] Bruce Schneier, *A Hacker's Mind: How the Powerful Bend Society's Rules, and How to Bend Them Back* (New York: W. W. Norton & Company, 2023).

High standards are what separate companies that say they care about quality from those that actually deliver it. It's easy to talk about excellence, but it's much harder to demand it every day, in every detail, from every person. That's why the best organizations don't just assume things will be done right. They check. They verify. They fix mistakes quickly and permanently.

This mindset shows up in everything from Amazon's hiring process to how it approaches security. It's why Andy Jassy threw someone out of a meeting over a missing Oxford comma. It wasn't about punctuation—it was about discipline. If you miss the small things, what else are you missing?

Insist on the Highest Standards is about building systems that guarantee quality. One of the most effective? The Andon Cord is a concept Amazon borrowed from Toyota, where everyone participates in quality control.

At Toyota, the Andon Cord hangs above the production line, and any worker can pull it the moment they spot a problem. Not to slow things down—to *stop* the entire line. At first glance, this might seem extreme, but that's the point. It prevents small defects from turning into bigger failures. The real power of the system isn't just the cord itself—it's the culture behind it. Employees aren't just allowed to stop production; they're *expected* to.

Naturally, American companies saw this and wanted in. If Toyota could build some of the highest quality cars in the world this way, surely this could help them too. Amazon and Netflix successfully adopted it, extending the concept beyond manufacturing into software, customer service, and product development. See a problem? Call it out. Fix it immediately. Don't let issues pile up. But these companies were successful because they built quality into their processes. The Andon Cord was part of a larger focus on quality.

Other companies decided to implement the Andon Cord as a one-off. But here's the problem: if you drop an Andon Cord into the wrong culture, it doesn't drive quality. It just creates chaos. Without the right foundation, the Andon Cord leads to constant disruptions. Work grinds to a halt. Leadership gets frustrated. Employees hesitate to pull it because they know it's just going to cause problems. Eventually, the whole thing gets scrapped because it "doesn't work here."

Everyone likes the idea of high standards until they realize what it actually takes. It's not just about setting a high bar—it's about refusing to lower it, even when it's inconvenient, even when no one's looking, even when it would be easier to let something slide.

LP #8

THINK BIG

Thinking small is a self-fulfilling prophecy. Leaders create and communicate a bold direction that inspires results. They think differently and look around corners for ways to serve customers.

When I told my boss, David Shaw, that I was going to do this thing, I went on a long walk with him in Central Park, and he said finally, after a lot of listening, "You know what, Jeff, this is a really good idea. I think you're onto a good idea here, but this would be a better idea for somebody who didn't already have a good job." That actually made so much sense to me, and he convinced me to think about it for two days before making a final decision. It was one of those decisions that I made with my heart and not my head, not wanting to pass up a great opportunity. When I'm eighty, I want to have minimized the number of regrets that I have in my life, and most of our regrets are acts of omission, things we didn't try, the path untraveled. Those are the things that haunt us.

—Jeff, Economic Club of Washington interview, 2018

When people hear Think Big, they imagine Amazon as the fairy godmother from *Cinderella*—flying around, granting wishes: "Little Amazonians, if you can dream it, you can do it!" Then, with a flick of her wrist, she creates new products and businesses.

I saw this magical thinking firsthand when I was head of banking at AWS. Banks would come to Amazon enthralled by Amazon Go, the cashier-less convenience store. Executives from around the world would ask me, "How can we build something as innovative as an Amazon Go store? We don't want Amazon Go exactly, but we want something just as groundbreaking."

Amazon is very open about what it takes to create something like Amazon Go. Just use the *Working Backwards* process from **LP #1: Customer Obsession**—then be willing to fail repeatedly for many years. Stay firm on your long-term goal but flexible on the tactics to get there. The game plan is simple, but it requires an enormous amount of work and dedication.

This is frustrating for many businesses. They want the results of innovation without being willing to accept failure. When I look at some of Amazon's biggest businesses—AWS, Kindle, Prime—it's hard to fathom how a company that started as a bookstore built them. Someone had to envision these massive ventures and work tirelessly on what might have seemed like a fool's errand. One of the original members on the Kindle said that it felt like they were either "getting one of the best seats on a rocket ship or working for years on a small business that never gets off the ground."[55]

Think Big is about three things. First, asking an absurdly large "What if?" question that seems insane at the time. "What if we could rent out data centers to other people?" Or "What if we could make everyone's living room a video store?" Then it's about creating a plan to get there. As my friend Nick says, "If you have an idea that works on paper, there's a small chance it could work. If you can't get it to work on paper, there's zero chance it will work." Finally, once you have the

---

[55] Colin Bryar and Bill Carr, *Working Backwards*, p. 163.

long-term vision nailed down, it's being flexible on the details and overcoming the many challenges to get there.

I remember when my friend interviewed at Amazon in 2011. The interview question was, "How can we get people to use the Amazon app on their phone to buy everything? We don't want them to think of buying anywhere else."

At the time, this seemed impossible. Everyone was still shopping for the best prices on Google and then purchasing from the retailer offering the lowest price. "What an absurd idea!" I thought. "No one would just buy everything on Amazon."

But today, for most things, Amazon is the one-stop shop. Looking back, 2011 seems almost quaint. How did Amazon change our buying habits so drastically?

Amazon Prime

The answer to that interview question began seven years earlier. In 2004, Amazon was starting to lose some of its luster. It primarily sold books and DVDs, and while the business was growing, its growth rate had slowed significantly. It was under attack from far larger brick-and-mortar rivals like Best Buy and Barnes & Noble.[56]

While Amazon had a better selection and great prices, customers hated paying for shipping—it felt like throwing money away. Free shipping was Amazon's most popular promotion, and customers valued it far more than the same amount of cash back.

Jeff had been pushing the team to explore different shipping alternatives. Then, in mid-October 2004, he sent an email to several executives, which said:

> We should not be satisfied with the growth of our retail business. This is a house-on-fire issue, and we need to dramatically

---

[56] Much of what follows about Amazon Prime comes from Vox's The making of Amazon Prime (audio).

improve the customer experience around shipping. We need a
shipping membership program. Let's build and launch it by the
end of the year.[57]

This was the critical time of year for Amazon, when everyone was
gearing up for "Peak," its term for the Christmas buying season. Jeff
asked the team to discuss the ideas behind Prime (then code-named
Futurama) on a Friday afternoon.

That Friday, however, the website suffered a multi-hour outage. The
team told Jeff they couldn't make the meeting due to the production
issue. Jeff responded, "Of course I understand. But this is so important
that you have to come over to my house on Saturday morning." At the
Saturday meeting, Jeff told the team, "I want to draw a moat around
our best customers. We're not going to take our best customers for
granted." He challenged them to create a shopping experience so
compelling that customers would want to buy almost everything from
Amazon—where it would be easier and cheaper (including shipping)
than their neighborhood store.

The team eventually settled on the free two-day shipping model
that became Amazon Prime. But that was just the beginning of the
story. The real challenge was making it work. The initial financial
models were terrifying. People love free shipping, and the first Prime
customers were the ones who shipped the most.[58] As Jeff later said,
"It cost us a lot of money because what happens when you offer a free
all-you-can-eat buffet? Who shows up to the buffet first? The heavy
eaters. It's scary. It's like, oh my god, did I really say as many prawns
as you can eat?!"[59]

To make Prime viable, Amazon needed to fundamentally change
its delivery system. Marc Onetto, former Amazon SVP of Worldwide

---

[57] Colin Bryar and Bill Carr, *Working Backwards*, p. 188.

[58] Amazon was charging $79 a year, which got you free two-day shipping and $3.99 for overnight shipping. The regular pricing of these things was $9.48 for two-day shipping and $16.48 for overnight.

[59] Jeff Bezos, Economic Club of Washington, September 13, 2018.

Operations, warned Jeff that the cost would be astronomical. At the time, air shipping cost $15 per package, while ground shipping cost just $1.50. Jeff responded, "You aren't thinking correctly." If customers loved Prime, demand would rise, giving Amazon the flexibility to build more fulfillment centers and drive down costs.[60]

Over time, Amazon made countless tweaks to optimize the system. They located more fulfillment centers closer to customers, reducing the need for costly air shipping. They limited two-day shipping for small items unless they were part of a larger order. For bulky items, free shipping was still available, but not necessarily in two days.

Eventually, more things came under the Prime umbrella. Prime Video and Prime Music were bundled in. Other services, like Amazon Fresh, the grocery service that had once cost $14.99 per month, became free for Prime members. Now, with over 200 million members, Prime has become a powerhouse for Amazon, generating more than $20 billion in revenue.

Looking back at that 2011 interview question, the strategy behind Prime was clear: Amazon wanted to be the first and only stop for online shopping. It all started with a challenge from Jeff. But it wasn't just about the Think Big vision—it was about execution. Every piece had to come together over time to make Prime successful.

As one journalist put it, "Amazon single-handedly—and permanently—raised the bar for convenience in online shopping. That, in turn, forever changed the types of products shoppers were willing to buy online."[61]

Amazon Invests in Risks and Risk Takers

Risk-taking at Amazon isn't just lip service. It actively rewards risk-takers. Just look at who Jeff tapped as his successor as CEO—Andy Jassy. Before leading Amazon, Jassy was the CEO of AWS, which, at the

---

[60] Jason Del Rey, "The making of Amazon Prime," Vox.

[61] Jason Del Rey, "The making of Amazon Prime."

time, accounted for only 12% of Amazon's total revenue. The obvious successor should have been Jeff Wilke, the head of Amazon's worldwide consumer business. However, Jeff chose Andy because he knew how to take a business from zero to a multi-billion-dollar enterprise.

Andy was one of Jeff's early Technical Advisors, informally called Jeff's "shadow." The shadow follows Jeff around to every meeting taking notes, is his primary sounding board for new ideas, and even drafts the annual shareholder letter. You might assume this role is a stepping stone to running one of Amazon's big existing businesses—but more often, it leads to building something entirely new. Andy wasn't the only one. Amit Agarwal went on to launch Amazon's business in India, and Dilip Kumar spearheaded the creation of the "Just Walk Out" convenience store Amazon Go.

Why is Amazon so focused on taking these risks? Jeff describes Amazon's innovation process like this:

> Most large organizations embrace the idea of invention but are not willing to suffer the string of failed experiments necessary to get there. Outsized returns often come from betting against conventional wisdom, and conventional wisdom is usually right. Given a ten percent chance of a 100 times payoff, you should take that bet every time. But you're still going to be wrong nine times out of ten. We all know that if you swing for the fences, you're going to strike out a lot, but you're also going to hit some home runs. The difference between baseball and business, however, is that baseball has a truncated outcome distribution. When you swing, no matter how well you connect with the ball, the most runs you can get is four. In business, every once in a while, when you step up to the plate, you can score 1,000 runs. This long-tailed distribution of returns is why it's important to be bold. Big winners pay for so many experiments.
>
> —Amazon 2015 Letter to Shareholders

Hunting for Black Swans

This pursuit of outsized returns was inspired by one of Jeff's favorite books: *The Black Swan* by Nassim Nicholas Taleb. He even had his entire Senior Leadership Team read it.

The title comes from an ancient Roman quote, "as rare as a black swan," which was the equivalent of "when pigs fly." At the time, no one in Europe had ever seen a black swan, so they assumed they didn't exist. As it turns out, there were black swans swimming around Western Australia at the time, only to be discovered by Europeans a millennia later. Black swans existed, but people couldn't conceive of them.[62]

In the book, Taleb defined the term this way:

> **Black Swan Event:** First, it is an outlier, as it lies outside the realm of regular expectations, because nothing in the past can convincingly point to its possibility. Second, it carries an extreme impact. Third, in spite of its outlier status, human nature makes us concoct explanations for its occurrence after the fact, making it explainable and predictable.[63]

For example, take the 2008 financial crisis. People knew the financial system was on shaky ground. A lawyer friend of mine, who worked on mortgage contracts back in 2004, told me, "All of these things depend on housing prices always going up. If they stop rising, really bad things will happen." Everyone knew the risk existed, but they ignored it. It was as if they were saying, "Housing prices can't go down, or horrible things will happen—so let's not even think about it."

While black swans can lead to catastrophes for those who ignore them, they also open up huge opportunities for those willing to go

[62] Currently we haven't found any animal that would be equivalent to a flying pig, but that doesn't mean there isn't one. Think about "flying squirrels" or a slightly different version of a bat.

[63] Nassim Nicholas Taleb, "The Black Swan: Chapter 1: The Impact of the Highly Improbable," *The New York Times*, April 22, 2007, https://www.nytimes.com/2007/04/22/books/chapters/0422-1st-tale.html.

against conventional wisdom. These moments have become classic business school case studies—cautionary tales repeated again and again.

Take Amazon's early days in the online book market. Borders didn't even put up a fight—it outsourced its online store to Amazon. Or consider Blockbuster, which passed on the chance to acquire Netflix because it was too committed to renting physical DVDs in brick-and-mortar stores. Then there's Kodak, which actually invented the digital camera but failed to embrace it because it was too reliant on film sales.

Taleb himself is a living example of a black swan success story. To understand Taleb, it's useful to compare him to a typical Wall Street trader—whom we'll call Mr. Wall Street.

As an options trader, Mr. Wall Street acts like a bookie in financial markets. If someone wants to place a bet, he sets the price and takes the other side. For every transaction, he collects a small cut—his vig, the gambler's take on the bet. Playing the role of a financial casino, he profits by absorbing lots of small risks over time.

On most days, Mr. Wall Street does quite well. It doesn't matter who wins each individual bet because, in the long run, things even out, and Mr. Wall Street still gets his cut. This is how he affords his penthouse on Park Avenue and his 13-acre compound in Greenwich.[64]

Taleb, however, plays a different game. He doesn't walk into Mr. Wall Street's casino to make small, incremental bets. Instead, he bets on a black swan event happening. He doesn't need to predict the exact event—he just makes a lot of large, cheap bets on things that no one else thinks will ever occur.

For example, he might bet that the stock market will lose 10% of its value in a single day. Mr. Wall Street happily sells him this option for a low price, convinced it's a waste of money.

Taleb loses money almost every day. He places bets, and they don't pay off.

---

[64] This is all from Malcolm Gladwell's New Yorker story "Blowing Up" in April 2002; Mr. Wall Street is based on Victor Niederhoffer.

It's hard to be Taleb. For years he will sit around losing money, day after day giving more money to Mr. Wall Street.

"We cannot blow up, we can only bleed to death," Taleb says.

This is hard to stomach. Humans aren't wired for this kind of patience. We expect immediate feedback. If we make money today, we assume we'll make money tomorrow. We don't expect to lose money every single day while waiting for an extremely rare event.

In a normal company, Taleb would get a terrible performance review—month after month of losses. But Taleb doesn't need a performance review anymore.

When a rare, unpredictable event does occur, Taleb cleans up.

He hasn't needed a regular job since 1987. On October 19, 1987—*Black Monday*—the stock market lost 22% of its value in a single day. Taleb made $35 million and never had to work a desk job again.

Since then, he has continued trading and has profited massively during other crises, including the turmoil after September 11 and the financial meltdown of 2008.

Final Thoughts

Think Big isn't just about having bold ideas—it's about staying with them long enough to see them through. It's about making bets that seem foolish, that drag on for years without clear results, and that most people would abandon long before they ever pay off.

Amazon didn't build AWS, Prime, or Kindle by playing it safe. It took on risks that seemed excessive, even reckless, at the time. It failed, recalibrated, and kept going until those ideas weren't just viable—they changed entire industries. The same goes for Taleb's black swan strategy. He spent years losing money, making bets that looked absurd—until one extraordinary event made it all worthwhile.

That's the reality of Think Big. It's messy. It's frustrating. It often looks like a series of bad decisions strung together—until, suddenly, it works.

And it's not just about business. The biggest things in life—relationships, careers, opportunities—follow the same unpredictable path. You put yourself out there, you try, you fail, and you try again.

We all have experience with Think Big when looking for a life partner. You sit through awkward first dates, make small talk that leads nowhere, and leave parties wondering why you even bothered. But you keep going, because it only takes one right moment to change everything.

That moment for me came 20 years ago at a party where I met my future wife. The funny thing is, she'd had a rough day and almost didn't come. But then she remembered her mom's advice: "Never skip a party invitation; the next one may change your life."

That night, it did. She showed up, and so did I. We met, we clicked, and we've been together ever since.

LP #9

BIAS FOR ACTION

Speed matters in business. Many decisions and actions are reversible and do not need extensive study. We value calculated risk taking.

One area where I think we are especially distinctive is failure. I believe we are the best place in the world to fail (we have plenty of practice!), and failure and invention are inseparable twins. To invent you have to experiment, and if you know in advance that it's going to work, it's not an experiment. Most large organizations embrace the idea of invention but are not willing to suffer the string of failed experiments necessary to get there.

—Jeff, 2015 Amazon Letter to Shareholders

In November 2019, Jeff received one of the greatest honors an American can receive: his portrait was installed in the Smithsonian National Portrait Gallery in Washington, D.C. He began his speech with, "My life is based on a large series of mistakes. I'm kind of famous for it in the business realm. How many people here have a Fire Phone? Yeah, no, none of you do... Thanks." It was classic Jeff: owning the failure, finding the humor, and making a point all at once.[65]

Amazon created the Fire Phone for the same reason it created the Kindle: to connect its content directly to users. At the time, Apple was trying to dominate the eBook market with iBooks, working with major publishers to ensure that they could always match or undercut Amazon's prices.

But Amazon was late to the game. By the time the Fire Phone launched in 2014, Apple was already on the iPhone 4. To stand out, the Fire Phone introduced features like Dynamic Perspective, which used four cameras to create a 3D-like effect, and Firefly, a tool that let users scan and identify items for purchase on Amazon.

The project was grueling. The team faced relentless demands and direct oversight from Jeff. Working on the Fire Phone wasn't just about the technical challenge—it was about meeting Jeff's expectations. "This was Jeff's baby," said one team leader.[66]

On June 18, 2014, Jeff launched the Fire Phone at Seattle's Fremont Theater. The product flopped. Hard. One of its biggest selling points—its deep integration with Amazon—ended up alienating users who wanted a more open system. Its standout features weren't strong enough to get people to switch from their iPhones and Androids.

In the short term, this was a massive failure. Amazon was left with a warehouse full of unsold Fire Phones and eventually took a $170 million write-down. But the lessons learned from the project shaped the company's future successes in consumer electronics. The Fire Phone's failures helped pave the way for later hits like Alexa. And the Amazonians who worked on it, through their painful mistakes, gained the skills they needed to build those future successes.

---

[65] Brad Stone, *Amazon Unbound: Jeff Bezos and the Invention of a Global Empire* (New York: Simon & Schuster, Kindle Edition, 2021), pp. 1–2.

[66] Austin Carr, "The Inside Story of Jeff Bezos's Fire Phone Debacle," Fast Company, January 6, 2015, https://www.fastcompany.com/3039887/under-fire.

Intelligent Failures

The Fire Phone was what Professor Amy Edmonson calls an "intelligent failure." In her book *The Right Kind of Wrong*, she describes two kinds of failure. "Basic failure" refers to avoidable mistakes caused by inattention or lack of knowledge. These are the missteps we instinctively try to prevent. "Intelligent failure," however, occurs in uncharted territory, where there's no clear path forward without testing the waters. These failures aren't careless; they're structured experiments, designed to generate new insights. They're the right kind of wrong.[67]

When we think of failure, we tend to focus on the first kind—the kind that feels wasteful, embarrassing, or preventable. The word itself often brings to mind carelessness, poor judgment, or a lack of preparation. But not all failures are created equal. Some aren't just useful—they're essential. In situations where there's no established playbook, the only way forward is to experiment. You try something, observe what happens, and adjust accordingly. It's messy, yes—but it's also how progress happens.

To illustrate this, Edmondson leads her Harvard Business School students through the Electric Maze exercise, a deceptively simple team challenge. Participants navigate a grid-patterned rug, searching for a path of non-beeping squares. There's no visible pattern, no prior knowledge to rely on—just trial, error, and the gradual process of mapping the way forward. The rules are simple: teams strategize briefly before starting, but once in motion, they must remain silent. Each time someone steps on a beeping square, they must start over. The team's success depends on their ability to accumulate and apply what they've learned.

On paper, it sounds easy. No specialized skills are required, no complex reasoning—just a willingness to test, fail, and remember. And yet most teams fail to complete the maze within the allotted 20

---

[67] Amy C. Edmondson, *The Right Kind of Wrong: The Science of Failing Well* (New York: Atria Books, 2023).

minutes. The reason? Hesitation. Faced with uncertainty, participants struggle to make the "right" decisions, hoping to pick the right next move instead of just testing their way forward. Their teammates encourage this behavior—cheering for successes and groaning at failures.

The exercise reveals a deep-seated instinct: we resist failure, even when it's necessary for success. A beeping square and a silent one offer the same thing—information—but our brains don't register them the same way. Progress feels good; missteps feel like we're doing something wrong. In reality, the quickest way through the maze isn't cautious planning or trying to guess the right move—it's brute force systematic testing. The teams that embrace this, stepping forward without hesitation, can solve the puzzle in under seven minutes. Most don't. They see the maze as a test to pass rather than a challenge to explore and can't finish it in the 20-minute time limit.

Of course, it's one thing to fumble through a maze in a classroom—it's another to make real decisions when the stakes are high. The instinct to slow down, gather more data, and escalate decisions upward makes sense. It feels safer. At a company like Amazon, a wrong call can cost millions. But speed isn't the real issue. The real discomfort comes from letting go of control—trusting others to make decisions, even when those decisions carry risk. And yet, that's exactly what's required. In complex, fast-moving environments, the people closest to the ground often have the best information. Driving decision-making downward may feel risky, but it leads to better decisions. A striking example of this comes from an unexpected place: the U.S. Army.

General Stanley McChrystal, former leader of Joint Special Operations Command (JSOC), realized that empowering his lower-level troops was key to effectively fighting terrorist networks. When McChrystal first arrived in Afghanistan, he had to personally approve all special forces missions. If the team needed to adjust their strategy based on new intelligence, they had to wake him in the middle of the night to get sign-off.

He wasn't in the best position to make those calls—he was exhausted, removed from the action, and lacked the situational awareness of the field commanders. Even if he could make a good decision, the time it took to get his approval often cost the team their window of opportunity.[68]

There was also the issue of accountability. If the mission failed, the commanding officer could go home and tell his wife, "General McChrystal had a bad day today. He made a really bad decision on that mission." The blame was entirely off the officer.

McChrystal changed the system. Instead of requiring approval, he told his teams: "Make the decision I would have made, given the information you have."[69]

One-Way and Two-Way Doors

Jeff wanted to empower Amazonians in a similar way. He recognized that employees often have more current, relevant information than senior leadership—and that many decisions are best made by those closest to the action. To support this kind of decentralized decision-making, he introduced the concept of one-way and two-way doors.[70]

One-way doors are big, irreversible decisions. They require careful, methodical decision-making because they will irreparably shape the company's future. Once you walk through this door, you can't get back to where you were before. Examples include Amazon's expansion into India or the acquisition of Whole Foods. These decisions need extensive discussion and planning and senior leadership attention.

But most decisions are two-way doors. Jeff says they should be made with about 70% of the information you wish you had. By that point, you have enough data to act. Gathering more information is costly and

[68] Stanley McChrystal, *Team of Teams: New Rules of Engagement for a Complex World* (New York: Portfolio, 2015).

[69] I heard this story from McChrystal when he gave a talk at Citigroup.

[70] Amazon Letters to Shareholders, 2015 and 2016.

time-consuming. It won't significantly improve your decision—and you'll likely need to correct your course anyway.

With a two-way door you can try something, go back, and then try something else. These are not decisions that should be made by committee. They should be made quickly by high-judgment individuals or small teams.

You Can't Grow Half a Beard

We're often afraid to commit to a specific path, opting instead for small, tentative steps toward a goal. But some things can't be done halfway. You can't do half of a skydive or ride half of a roller coaster. Yet people still believe they can cautiously dip a toe into a new idea and see if it works.

This really hit me when I was at Disney World. We were eating dinner at the 50's Prime Time Café in Hollywood Studios, being served by a waitress playing the role of "Mom," who constantly reminded us to mind our manners and clean our plates. At one point, I noticed a skin-colored fabric sleeve on her bicep.

"What's that?" I asked.

"Disney is very particular on how we dress inside the park," she said. "They've gotten more flexible recently. I've got a tattoo, but management is fine if I cover it. They've also changed a long-standing policy and now allow beards, as long as they look neat. As soon as that happened, a bunch of guys took a week of vacation."

"Why?"

"They needed time to grow out their beards so they looked neat. They knew they couldn't show up to work with a half-grown beard."

How We Overestimate WTF Will Happen

We often spend an enormous amount of time trying to design the perfect solution upfront—as if we can fix something once and for all.

But Jeff knows better. He says, "Customers are always beautifully, wonderfully dissatisfied, even when they report being happy and business is great." That dissatisfaction isn't a bug; it's part of the system. The moment you think you've solved a problem, a new one quietly takes its place. It's a never-ending game of whack-a-mole.

Tim O'Reilly explores this idea in his provocatively titled book *WTF*. He describes two types of "WTF" moments in technology. The first is the awe-struck kind: *"WTF?! I didn't even think that was possible!"* The second shows up a little later: *"WTF?! I can't believe this doesn't work better!"* We go from wonder to frustration faster than we'd like to admit.

1. WTF?! I didn't even think that was possible!

The first time you see Google Maps, you say something like, "WTF?! This is amazing! I can get a map of the entire world and directions to anywhere I want to go!" Or, decades ago, "WTF?! FedEx will deliver my documents in two days anywhere in the world!"

2. WTF?! I can't believe this doesn't work better!

After a while, that same breakthrough feels outdated. "WTF?! Google Maps keeps sending me into traffic! It's useless without an internet connection!" Or, "FedEx is fine, but why does it take so long?! Amazon delivers in a few hours and gives me much better status reports."

This is why decisions need to be treated as experiments. No matter how much planning or analysis you do upfront, you'll still have to adjust. And the more certain you are that the plan is right, the harder it becomes to change course when it's not.

I once worked at a company where a product manager was spending a lot of time building the next version of his product. He believed he could solve a customer problem once and for all with enough planning. Our conversation went like this:

"Why is this version taking so long?" I said.

"Solving this problem properly takes time. I'm incorporating everything we've heard from users—every piece of feedback, every issue—and making sure we get it right this time," he said.

"If you get this done, will the problem be solved?" I said.

"Yes," he said.

"So will you still have a job?" I said.

"Of course," he said.

"But if the problem is solved, why would you still have a job?" I said.

That's the flaw in the thinking. Product managers are there to manage the changes that are going to happen. The market shifts, competitors make moves, and customer demands evolve. So don't spent too much time solving today's problems. The problem you started with won't be the problem you end with.

I once saw a company spend two years debating how to respond to social media—just as the trend was gaining momentum. They believed they were being cautious, but in reality, they were stuck in analysis paralysis. By the time they finally made a decision, the first wave of social media had already passed. Instead of acknowledging that they'd moved too slowly, a senior leader said, "See? Wasn't it good we didn't act? We would've had to shift focus to a new trend anyway." It sounded reasonable, but really, it was just a way of justifying hesitation.

And yes, the early days of social media were a little ridiculous— virtual farming, poking people, endless zombie invites—but that wasn't the point. What they missed was the chance to learn how the space worked while the stakes were still low.

Embracing the Randomness of the World

When we focus too much on finding the perfect answer, we make an organization fragile. It can't evolve. It gets stuck in a mindset that assumes the world will remain predictable—even though it never does.

So what can we do instead?

After writing *The Black Swan*, Nassim Nicholas Taleb—our contrarian trader from the last chapter—expanded on his ideas with the book *Antifragile*.[71]

Being antifragile is more than just being tough. Fragile things break when they're shaken. Robust things don't break, but they don't improve either. Antifragile things, on the other hand, actually get stronger under stress.

It sounds counterintuitive, but most natural systems thrive when tested and challenged. Cities, cultures, ideas, even technological innovation—they all evolve by bumping up against the unexpected. Man-made systems tend to be fragile. Natural ones are antifragile.

Taleb opens *Antifragile* with a line that captures this perfectly:

> Wind extinguishes a candle and energizes fire. Likewise with randomness, uncertainty, chaos: you want to use them, not hide from them. You want to be the fire and wish for the wind.

Consider how evolution helps keep us healthy. No matter how advanced our antibiotics become, pathogens continue to evolve in unexpected ways. Our immune system acts like a security system with locked doors, and pathogens are constantly trying to find a way in. If the locks never change, the pathogens will eventually crack the code.[72]

The best defense against disease isn't just better drugs—it's changing the locks. We do this through sexual reproduction. By constantly mixing genetic material, nature creates new combinations that pathogens must continually adapt to. Some of these "locks" may be weaker than before, but others are stronger. It's this genetic randomness that helps keep us healthy as a species.

In business, though, we resist randomness. We strive for perfection and uniformity, trying to eliminate variation wherever possible. But the more we fight the natural world, the more vulnerable we become.

---

[71] Nassim Nicholas Taleb, *Antifragile: Things That Gain from Disorder* (New York: Random House, 2012).

[72] Steven Pinker, *How the Mind Works* (New York: W. W. Norton & Company, 1997).

This pursuit of perfection and the fight against disease are colliding in an unexpected place: the survival of the banana.

Have you ever noticed that every banana you eat tastes the same? That's because they are the same banana.[73] Nearly all bananas sold globally are a single variety—the Cavendish: firm, easy to ship, and relatively tasty. It is the world's best banana from an economic perspective. But there's one massive downside: the Cavendish reproduces asexually, meaning every single banana plant is genetically identical.

Right now, a fungus called Panama Disease TR4 is spreading across the world, targeting the Cavendish. Since every plant has the same genetic makeup, they all have the same vulnerabilities. The fungus doesn't just kill individual plants—it wipes out entire plantations. And because 99% of the $44 billion banana industry depends on this one fragile variety, the entire business is at risk.[74]

Final Thoughts

Bias for Action isn't about being reckless—it's about having the courage to move before everything is fully known. It means recognizing that in fast-changing environments, hesitation can be more dangerous than a wrong turn. Whether it's launching a product like the Fire Phone, navigating an unfamiliar maze, or fighting terrorists, the pattern is the same: progress comes through doing. We learn by trying, by adjusting, by moving forward—not by standing still and waiting for the perfect answer to arrive.

The world doesn't stop moving. Technologies shift, markets evolve, and customer expectations never really settle. In that kind of environment, standing still starts to look a lot like falling behind. The smarter move—the more resilient one—is to keep moving. If you

---

[73] Note that other parts of the world, like Singapore, have a much larger variety of bananas than I've eaten.

[74] Stuart Thompson, "The Quest to Save the Banana from Extinction," *The Conversation*, April 18, 2019, https://theconversation.com/the-quest-to-save-the-banana-from-extinction-112256.

treat decisions as experiments, give yourself room to adjust, and build systems that respond well to the unexpected, you don't just weather change—you learn from it. You get stronger because of it. Or, as Taleb might say, you become the fire that welcomes the wind.

It's about moving forward, even when you're not entirely sure where the path leads.

Back in college, I started writing for a national lifestyle magazine— not something I planned, but something I more or less stumbled into. An editor had come across a piece I'd written for my college humor magazine and asked if they could reprint it. One thing led to another, and before long I was listed as a Contributing Editor, writing a good chunk of each issue and landing increasingly ambitious assignments. One of them was a visit to George Lucas's Industrial Light & Magic and Skywalker Ranch.

For a certain kind of person (me), this was the pinnacle. Industrial Light & Magic—ILM—is the visual effects studio behind *Star Wars*, *Jurassic Park*, *Indiana Jones*, and just about every blockbuster you can name from the last few decades. Skywalker Ranch, tucked into the hills of Marin County, is Lucasfilm's private creative campus. It looks like a vineyard, but it's packed with soundstages, edit bays, and screening rooms. The whole place runs on a blend of craft, secrecy, and myth.

The trip was a dream assignment. But when I got back, I sat there, staring at my notes, completely overwhelmed. There was too much to say, and I had no idea how to start. I asked a few writer friends for advice, and someone told me a story about Tom Wolfe.

Wolfe was one of the big voices in American journalism—famous for *The Right Stuff* and *The Bonfire of the Vanities*—but early in his career, he got stuck on an article about custom cars. His editor, Byron Dobell, told him to forget the article and just write a letter explaining what he saw. Wolfe did exactly that—49 pages' worth. Dobell took one look, scratched out "Dear Byron," and published the whole thing as-is.[75]

---

[75] Michael Lewis, "How Tom Wolfe Became ... Tom Wolfe," *Vanity Fair*, October 8, 2015, https://www.vanityfair.com/culture/2015/10/how-tom-wolfe-became-tom-wolfe.

So I tried the same approach. I dumped everything I had onto the page—figuring the editors would cut it down. Instead, they ran the whole thing, three times the assigned length. And because I was paid by the word, they paid me three times as much.

LP #10

FRUGALITY

Accomplish more with less. Constraints breed resourcefulness, self-sufficiency, and invention. There are no extra points for growing headcount, budget size, or fixed expense.

I think frugality drives innovation, just like other constraints do. One of the only ways to get out of a tight box is to invent your way out.

—Jeff, "Bezos on Innovation" Bloomberg interview, 2008

When Jeff founded Amazon in 1994 in a small garage, the company needed desks. But they were strapped for cash, and desks were pricey. While walking through Home Depot, the team noticed that solid core doors were much cheaper. So they bought the doors, grabbed some 2x4s, and built their own makeshift desks. Thus was born the famous "Amazon door desks." For many years afterward, employees would sit at these makeshift desks to remind them of frugality. They even built their own desks as part of orientation.

The door desk symbolizes Amazon's focus on frugality. Frugality isn't about being cheap—it's about being thoughtful with resources.

When a budget forces trade-offs, employees prioritize what's truly needed. By being frugal on things like office furniture, the company could spend more on what mattered to customers—better service, better products, better prices. Staying frugal meant Amazon could deliver on its other core promise: offering the lowest prices.

By the time I joined Amazon, they had stopped using door desks, but frugality was still front and center. Amazon still didn't offer the standard perks of other Silicon Valley companies. There was no gourmet cafeteria, just a small corporate cafeteria to grab food if you were in a rush. We were told that Amazonians should explore the local neighborhood and support local businesses. Like many things about Amazon's frugality, it wasn't wrong. Most large companies, like banks, operated the same way. But compared to other tech companies, it didn't feel great.

I had a friend who worked at LinkedIn across the street from our office, in the Empire State Building. They had free lunch. And what a beautiful lunch it was. Their cafeteria seemed to take up an entire floor, serving entrees like roast chicken and steak alongside homemade soups and salads. When I asked about dessert, my friend told me the company only served healthy food—no sweets—but I might be able to find some leftover pastries from breakfast.

When I got back to the Amazon office, I told my coworkers about the culinary wonders across the street. I couldn't stop blathering. They stared at me wide-eyed, their brains short-circuiting. I could almost hear them thinking, "LinkedIn is a tech company. We work for a tech company. Why are these companies so different?"

Frugality wasn't just culture—it was policy. Amazon was well known for not paying for anything above coach class on flights. If Jeff flew on a private plane, it was Jeff's plane, not Amazon's. I wasn't sure this was really enforced—until I flew to India and Japan in coach. In true Amazon fashion, if a trip benefits the customer, you should go. But no one said you had to be comfortable getting there.

The difference between Amazon and other tech companies was summed up by one of my Amazonian friends that used to work at Google. He said, "The difference is that Google fetishizes its employees, while we fetishize our customers."

One example? The Amazon employee discount. When I started, everyone asked, "Do you get a discount?" I was excited to say yes—until I found out what it actually was. Employees got 10% off items sold directly by Amazon (not third-party sellers), capped at $100 a year. That's it. Meanwhile, customers with an Amazon Prime credit card got 5% back on everything—Amazon purchases, Whole Foods, the works—with no limit.

The message was loud and clear. If you wanted the best deal on Amazon, it was better to be a customer than an employee. And that was very on brand.

The Power of Less

I learned the most about frugality by focusing on small everyday moments. It started with fries. I noticed that when I'm dining out on an expense account, my entire mindset around food shifts. When the waitress asks, "Do you want fries with that?" or "Would you care for dessert?", the cost of that extra food drops to zero in my mind, making it much easier to say yes. I first noticed this early in my career as a consultant. I traveled Monday to Thursday, living off an expense account, and before I knew it, my weight started to balloon.

A few years later, I had the opposite experience. I wanted to lose weight but wasn't sure how. By then, I wasn't traveling and was eating lunch in our corporate cafeteria. At the salad bar, there were two container sizes: a "regular" 10-inch square container and a smaller 6-inch side salad container.

I had always used the regular container because that was what I was "supposed" to use—it was the standard size. And with 10 inches to fill, I acted like a wealthy aristocrat selecting ingredients for a grand feast. "A big spoonful of berries, some chicken, some tomatoes, a nice dollop of blue cheese dressing on top. And to make it healthy, I'll put it over a generous bed of lettuce." But even with all that food, I'd still need an afternoon snack to keep from getting hungry.

Then one day, I decided to try the 6-inch container instead. It completely changed the way I thought about food. Instead of a wealthy aristocrat, I became a curator of my miniature salad. I approached the tongs like a bonsai master. "Four pieces of melon. Three meatballs. A small amount of spaghetti to go with the meatballs. Oh, there's just enough room for five blueberries." I wasn't obsessive about it, but I appreciated each item so much more.

Frugality Is About Prioritization

Frugality is about being thoughtful with resources. It's not just about doing things right—it's about doing the right things. In school, we were taught that the goal was to score 100% on a test. The objectives were clear, and our job was to achieve them. But life doesn't work that way. There's no perfect answer. Everything we choose to do means choosing not to do something else.

As Steve Jobs said, "People think focus means saying yes to the thing you've got to focus on. But that's not what it means at all. It means saying no to the hundred other good ideas that there are."

I remember a conversation with my mentee, Rachel. She said, "Rob, what should I do? I'm working for three senior executives, and they all need their projects done this week. They're in different groups, and all the work is important. Normally I can handle it, but this week my days are so packed I physically can't do it."

First I gave her the standard advice: talk to her stakeholders to see what could be pushed to next week, and check if someone else could help. But then I told her, "If your days are regularly filled to 100%, you have a different problem. If your goal is to max out your time, you're focusing on staying busy. That's useful at a junior level, but as you get more senior, the game changes. It's not just about checking off tasks—it's about making space to decide which tasks are worth checking off."

To do this effectively, you need to do two things:

1. Capture all the things you might want to do.
2. Ruthlessly prioritize based on value.

Make Sure You Capture All of the Possibilities

The trick is to actively look for new possibilities and capture them. This sounds easy, but it requires a mindset shift. Here's the difference. As an overburdened product manager, when someone gives you an idea, your gut reaction might be, "Oh no, I have too much to do already. Here comes another thing. Maybe if I ignore it, it'll go away." But with the right mindset, when a well-thought-out idea comes your way, you say, "That's interesting. I have a lot on my plate, but this is worth considering—maybe not right now, but I'll keep it for later."

Ideas can come from anywhere, but the best ones often come from customers. These are the people in the weeds and doing the work, navigating inefficiencies, and figuring out how to get around them. Often they've come up with clever solutions that you can integrate into your product. However, some people are afraid of talking to customers about their products. I once had a colleague who said, "We can't ask them what they want. If we ask, they'll expect us to deliver." But the trick is to listen—let them tell you what they need—without committing to anything.

Some ideas will be fantastic and deserve immediate attention. If something feels like a clear win—high impact, low effort—don't overcomplicate it. Just try it. A quick test or lightweight prototype can validate whether it's worth building. But most ideas won't be that simple. They might be too expensive, too niche, or just not quite right for the moment. That doesn't mean they're bad. Often, the best thing you can do is set them aside and give them space. Let them marinate, evolve, or collide with something else later on. Ideas have a way of ripening over time—what doesn't work today might be exactly what you need a year from now.

Ruthlessly Prioritize Based on Value

Once you've gathered all the possibilities, the hard part isn't figuring out how to start—it's deciding what not to do. There's a certain comfort in the idea that more is better, that if you just work hard enough, you can get through the entire list. But at some point, usually sooner than we'd like, reality steps in. Time, energy, and attention are finite. Something has to give.

Marie Kondo's advice about clutter comes to mind here. She's known for saying, "If it doesn't bring you joy, throw it out." In a business setting, the language shifts, but the principle holds. If it doesn't create real value—especially for the customer—it's probably not worth doing.

I saw this play out while helping my friend Catherine edit a speech. She had a five-minute version that needed to be cut down to a pithy one minute. At first, it seemed impossible. There was nothing that was bad—nothing obvious to cut. But then I started looking at the speech with a different lens. What was truly amazing versus just good? I kept the strongest pieces and cut everything else. The speech wasn't just shorter; it was a lot more powerful. Keeping only the best material and removing the "good but not great" content allowed the average quality of the speech to skyrocket.

Once you have your long list of captured ideas, you need a way to prioritize them. One particularly useful framework comes from Agile development:

1. Estimate the **impact** of each item.
2. Estimate the **effort** that each item will take.
3. Divide the impact by effort to get the **value** of each item.

4. Order the list based on value.
5. Move the priorities based on real-world considerations. Is there a specific reason something needs to be done today? Is someone waiting for you and really needs something?
6. Move down the list one at a time until you run out of capacity.

I used to think that if I just had the right process, we could get all the right things done. That there was some magic formula for prioritizing—something that would help me see clearly what should come first. But when I was leading a global product, I realized it was less about getting it right and more about letting people own their priorities. Each stakeholder had its own laundry list of requests, always more than we could realistically deliver.

My predecessor tried to solve this by making the decisions himself—choosing what he thought would be most useful across the board. It didn't go well. The stakeholders were constantly arguing that their priorities were the most important.

Then I realized something very simple. Instead of prioritizing the work myself, I could have the individual stakeholders do it. I dedicated a fixed amount of capacity to each stakeholder and asked them to decide how they wanted to use it. Once they had ownership, they focused less on fighting for more and instead tried to make the most of what they had.

One of my favorite examples of ruthless prioritization comes from Joel Spolsky, a software entrepreneur who knows a thing or two about building great products. Joel was an early product manager at Microsoft, wrote the influential *Joel on Software* blog in the early 2000s, and later founded Trello and Stack Overflow—making gobs and gobs of money. He once described the roadmap planning for Excel 5:

> I remember working on Excel 5. Our original feature list was huge and would have gone way over schedule. Oh my! we thought. Those are all super important features! How can we live without a macro editing wizard?

As it turns out, we had no choice, and we cut what we thought was "to the bone" to make the schedule. Everybody felt unhappy about the cuts. To assuage our feelings, we simply told ourselves that we weren't cutting the features, we were simply deferring them to Excel 6, since they were less important.

As Excel 5 was nearing completion, I started working on the Excel 6 spec with a colleague, Eric Michelman. We sat down to go through the list of "Excel 6" features that had been cut from the Excel 5 schedule. We were absolutely shocked to see that the list of cut features was the shoddiest list of features you could imagine. Not one of those features was worth doing. I don't think a single one of them was ever done, even in the next three releases. The process of culling features to fit a schedule was the best thing we could have done. If we hadn't done this, Excel 5 would have taken twice as long and included 50% useless crap features.[76]

Final Thoughts

It's easy to confuse frugality with being cheap. But at its best, it's not about doing less for the sake of less—it's about doing more of what matters by letting go of what doesn't. Whether it's building desks out of doors, skipping employee perks, or trimming a speech down to just the essential words, the heart of frugality is being intentional. It's the quiet discipline of focusing your time, money, and energy where they have the most impact.

Of course, that kind of clarity isn't always obvious in the moment. Saying no to something good can feel wrong—wasteful, even. But if everything's important, nothing really is. Whether you're writing a speech or planning the next version of Excel, the challenge is the same: separating the must-haves from the nice-to-haves. The goal isn't just brevity. It's focus. By removing what's merely good, you give

[76] Joel Spolsky, "Painless Software Schedules," *Joel on Software*, March 29, 2000, https://www.joelonsoftware.com/2000/03/29/painless-software-schedules/

the essential pieces space to land. You make the whole thing sharper, more deliberate. Less noise, more signal.

The same idea holds in the rest of life. Living frugally—or intentionally—isn't about self-denial. It's about being clear-eyed—about knowing what to focus on. In the end, our days are shaped by what we make time for. And that means returning, again and again, to the question of what really matters.

Sometimes, frugality isn't about self-denial at all. Sometimes, it's about letting go of the sense of obligation—the "should"—and choosing what's actually best in the moment.

At my college reunion, my friend Lutz and I had packed our day with activities. As we walked down the street, I saw a sign for Ashley's Homemade Ice Cream. I said, "Lutz, I'd really like to buy my family some ice cream."

Being the logical German, he replied, "Yes, that would be nice, but there's no time. We need to go hear the university president's speech."

"Lutz," I said, "you have to prioritize. You can always watch a video of the speech or read a transcript. But how often do you get to sit with your family on a beautiful spring day, enjoying ice cream on the college lawn?"

He thought for a second. "You're right. Let's get the ice cream."

And it was the best decision we made that weekend.

LP #11

EARN TRUST

Leaders listen attentively, speak candidly, and treat others respectfully. They are vocally self-critical, even when doing so is awkward or embarrassing. Leaders do not believe their or their team's body odor smells of perfume. They benchmark themselves and their teams against the best.

A brand for a company is like a reputation for a person. You earn reputation by trying to do hard things well.

—Jeff, "Online Extra: Jeff Bezos on Word-of-Mouth Power" Bloomberg interview, 2004

How does Amazon build trust? By being helpful and honest. By shopping with Amazon, customers quickly learn they can get exactly what they want, cheaply and fast. They trust that the company will not just deliver their goods, but also deliver on its promises. Over time, customers see that Amazon is serious about customer obsession, continually improving the customer experience year after year.

Amazon is honest with customers, even when it isn't easy or convenient. That kind of candor can feel risky in the moment, but it's also what sets the company apart. By staying focused on long-term trust and true customer obsession, Amazon has often made decisions that didn't serve its short-term interests—but built something stronger over time. One famous example from Amazon lore appears in the 2003 Letter to Shareholders. Jeff wrote:

> Shortly after launching Amazon.com in 1995, we empowered customers to review products. While now a routine Amazon.com practice, at the time we received complaints from a few vendors, basically wondering if we understood our business: "You make money when you sell things—why would you allow negative reviews on your website?" Speaking as a focus group of one, I know I've sometimes changed my mind before making purchases on Amazon.com as a result of negative or lukewarm customer reviews. Though negative reviews cost us some sales in the short term, helping customers make better purchase decisions ultimately pays off for the company.[77]

Even in 2024, Amazon continues to be honest, even when it hurts. When I was at Amazon, there was a big push to have customers use microservices (see **LP #2: Ownership**). The idea was that it was better to slice your application into smaller and smaller pieces. This allowed the system to run more efficiently using some sophisticated AWS technologies. At least that was the theory. Listening to its own advice, Amazon split its Prime Video service into many pieces. Then, in 2024, the Prime Video team published an article (since deleted) that shocked many people in the tech community. It said that Amazon was getting rid of microservices for Prime Video. For this application,

---

[77] Jeff Bezos, "2003 Letter to Shareholders," Amazon.com, accessed March 30, 2025, https://www.amazon.com/ir/shareholder-letter-2003.

splitting it into all those tiny pieces wasn't the right thing to do. According to the article, it was 10 times more expensive due to the operational overhead.[78]

Looking at that article, it's easy to think, "Who let that team publish this? This flies in the face of all the theories we've been preaching for years." How could Amazon be so honest? Isn't it nervous about what customers might say? But this honesty is why customers trust Amazon.

This trust and loyalty are built on having the best product and providing the best advice. One of the unique aspects of Amazon is that it doesn't use its market power for quid pro quo deals. Normally, here's what happens: When I was at a large company, we pitched a product to a very large customer. We had a better product and thought we had a chance. However, a former executive from the customer told me, "We don't have a shot. The customer spreads their money around based on who buys from them, and since the competitor only sells one product, that's the one the customer will buy. There's no way we can win."

Amazon doesn't operate that way. Amazon doesn't say to Chase, "Look, you're our partner for our Amazon credit card. If you don't use AWS, we're going to look for another partner." As much as the AWS team would have liked that, it never happened. Not even close. To win a customer, you had to build the best products.

Amazon prefers to use internal tools, but if you have a business need, you can use something else. Amazon sells a conferencing tool called Chime that competes with Zoom and Teams. It was widely used by Amazonians for internal and external meetings. However, few people outside Amazon had it on their computers. The private joke was that if I saw someone using Chime on a plane, I knew they worked for Amazon, even though it was a public tool—I was never wrong. But if you were a salesperson, you didn't have to use Chime. You could easily request

---

[78] "Scaling Up the Prime Video Audio/Video Monitoring Service and Reducing Costs by 90%," Prime Video Tech, January 18, 2024, https://web.archive.org/web/20230609112905/https://www.primevideotech.com/video-streaming/scaling-up-the-prime-video-audio-video-monitoring-service-and-reducing-costs-by-90.

the tool of your choice if it made your job easier or if the customer preferred something else.

Building Trust Through Branding

A company's brand shapes how people see and trust it. Amazon's brand is not just about the orange logo you see on packages. It sets the standard for how the company will act and what customers can expect. That's why Jeff says, "A brand for a company is like a reputation for a person. You earn reputation by trying to do hard things well."

Trust in a brand, much like in personal relationships, is cultivated over time by the way the brand acts. A brand earns trust by consistently delivering on its promises, maintaining quality, and aligning its actions with its values. This consistency lets customers know they can rely on the brand to meet their expectations, which is the heart of long-term loyalty.

A brand is a magical thing. It creates something from nothing. It's an intangible concept that only exists because people trust and believe in what it represents. Take Coca-Cola. Strip away the branding, and you've got sugar, water, and bubbles. But when it's bottled into that red can with that logo on it, it becomes a moment of joy, a taste of nostalgia, and a connection to others. While it's easy to be cynical and say that Coca-Cola is convincing people that sugared water is happiness, it really is creating something wonderful that wasn't there before. It turns an ordinary soda into something transcendent.

The Coca-Cola brand goes far beyond television commercials and marketing slogans. Much like Amazon, it's embedded into the core of the company. When I visited the Coca-Cola headquarters for work in the early 2010s, every whiteboard had a small sign that said, "Have you offered your visitors a Coke?" This simple sign was a reminder of what this company is all about. *We're not here just to make money. We're not even here to sell Coke. We're here to share a Coke with others.*

Brands do more than change how products look or how employees act; they build whole ecosystems. For example, Marriott runs 30 different hotel brands to meet all kinds of customer needs. To understand how these brands influence customer behavior, I chatted with a hotel owner who told me about the worst customers at both high-end and budget hotels.

- **High-End Hotel:** The worst kind of customer has a sense of entitlement that borders on—and sometimes crosses into—the unreasonable. They nitpick the smallest details, demand instant solutions, and expect royal treatment at all times. This is the stereotypical "difficult customer"—the kind who treats "the customer is always right" as a license to say and do whatever they want.
- **Budget Hotel:** You might think the worst customer at a budget hotel would be easier to deal with, given their lower expectations. But here, the worst customer might come in drunk and start a fight with the manager. The hotel owner described one particularly disturbing fraud carried out by guests. Customers who already have bedbugs try to blame it on a hotel. They bring infested linens into their hotel room. After their stay, they leave, giving the bedbugs time to settle in. A few days later, these guests sue the hotel, claiming their home got bedbugs from their stay. Now the hotel faces not only a lawsuit but also an infestation.

Trusting Amazon with Your Company's Future

When a company is considering AWS for serious transformational projects, they are invited for a day of meetings at Amazon's Executive Briefing Center (EBC). At an EBC session, customers are invited to Amazon's mothership to get closer to the company. Here, they aren't just visitors—they are treated as VIPs. They are invited to explore the latest technology roadmaps, hear about Amazon's Leadership Principles, and visit unique spaces such as Amazon's internal botanical terrarium, The Spheres.

These meetings showcase the latest and greatest from AWS, offering customers a front-row seat to the company's future. They meet with product managers of the services they're exploring and learn what's coming next. This backstage pass brings customers along on the AWS journey, inviting them into a real partnership. Then the team shares success stories from similar projects, allowing the honored guests to envision where they will be in a few years.

These showcases might seem like just another way to drum up more business, but they're really about building a deep partnership grounded in mutual investment. Both sides are committing. Senior executives from the customer give up three days—two for travel, one for the EBC. AWS executives set aside time from their own packed schedules. The result? Customers walk away feeling like AWS has made a serious investment in them—they must matter.

The EBC also gives customers direct access to the people building AWS products. They can answer pointed questions like what trade-offs were considered during development, and whether a given service can really scale to their demands or if it still needs some time to mature. They learn that things that seem simple on the surface are often deeply complex underneath. By the end, many conclude, "We can't hire this level of talent to build this ourselves. We need a partner—and none of our vendors give us this kind of access. The best way to get to the future is with AWS."

None of this happens without trust. The decision-maker—CIO, CTO—isn't just trusting AWS with their company's future. They're trusting AWS with their own career. That leader walks into a boardroom and essentially drops their badge on the table: "If AWS doesn't work, you can fire me." What does it take to earn that kind of trust? An EBC— or two, plus an annual refresher. I remember one CEO of a large bank who traveled with his executive team halfway around the world for an EBC. At the end of the day, he said, "Today wasn't a meeting about technology. It wasn't even a meeting about the future of our bank. Today's conversation was extraordinary and possibly life-changing."

The Front Door to Everything

Amazon is one of the world's most trusted brands, and that trust has enabled it to expand into areas like grocery, filmmaking, and consumer electronics. This reputation makes Amazon the go-to destination for nearly any purchase. Over time, Amazon has transformed the retail experience. To better understand that shift, consider Airbnb, which similarly reimagined lodging.

Travelers have always had the same basic needs. They want a place to stay that's comfortable, safe, and close to their activities. Before Airbnb, hotels owned their buildings or operated through franchisees, delivering consistent experiences defined by corporate standards. A Marriott Courtyard in Tokyo is the same as a Marriott Courtyard in Boise.

Airbnb, by contrast, sets traveler expectations during the booking process. Instead of offering uniform standards like bed type, free breakfast, or complimentary Wi-Fi, it focuses on the individual experience of each rental. This lets Airbnb commoditize its supply one room at a time while still maintaining a cohesive customer experience. Airbnb becomes the only brand that matters in this value chain—and that makes it the most valuable.[79]

Amazon plays a similar role as the front door of the world's largest marketplace. In addition to selling its own products, it provides a platform where third-party sellers can reach millions of customers. By aggregating these sellers into a single superstore, Amazon offers an unmatched selection that no single retailer could achieve on its own. It effectively commoditizes supply, making the sellers interchangeable

[79] Ben Thompson, "Airbnb and the Internet Revolution," *Stratechery*, July 1, 2015, https://stratechery.com/2015/airbnb-and-the-internet-revolution/.

while keeping customer loyalty centered on the Amazon brand—not the individual vendors.

But Amazon takes this a step further with its private label line, Amazon Basics. Just as it connects customers to third-party sellers, with Amazon Basics, it connects them directly to manufacturers. The Amazon brand guarantees quality and reliability, bridging the gap between producer and consumer without the traditional retail middleman. This not only builds trust in those products but also gives Amazon more control over the entire value chain.

Final Thoughts

Trust turns a company into something more than a collection of products or services. It turns it into a brand people believe in—not because they were told to, but because experience has taught them to. That belief is earned slowly, over time, through choices that are often harder in the moment but better in the long run.

Amazon doesn't get everything right. No company does. But what sets it apart is its instinct to own the hard parts, to speak plainly, and to keep showing up for customers—even when it's uncomfortable. Whether it's allowing negative reviews, publishing an article that challenges its own long-held architectural decisions, or declining to use its market power to pressure a partner, Amazon understands that trust can't be faked. You have to live it.

It's not flashy work. There are no shortcuts. But over time, that consistency builds something durable. It's what gives a customer the confidence to click "Buy Now" without thinking twice. It's what lets a CTO walk into a boardroom and say, "We're betting on AWS," and mean it. And it's what allows Amazon to keep trying new things, knowing customers will give it the benefit of the doubt—not because it's perfect, but because it's proven trustworthy.

That kind of trust doesn't only matter in business. It's personal, too. When I think about trust on a more human level, I think of my father. To me, trust was watching someone quietly do what needed to be done—no drama, no shortcuts.

He was working on a project overseeing TV broadcast equipment. One day, I visited him on the job as he managed the feed for a large sporting event. About 30 minutes before the event started, he did a final check. The live feed to the TV trucks was dead. "Dad," I asked, "what happens if you don't fix it?" He told me that no one would be able to watch the event. Calmly, step by step, he found the issue and fixed it with minutes to spare. That's why they trusted him. He always delivered.

That same sense of trust—earned through quiet reliability and openness—shows up in mentorship, too. A trustworthy mentor offers not only advice, but access—insight into how senior leaders think and operate. It's hard to understand what leaders are truly thinking or even to see them as real people. A mentor who trusts you enough to pull back the curtain is giving you a real gift. I remember one of my mentors once told me, "You don't have enough time to be an expert in everything. You do it a couple of times and then move on to something new." This mentor had been the CEO of several companies. Hearing that even he didn't always know everything changed how I thought about my career.

I've tried to pay that forward with my own mentees by being vulnerable and letting them know that I'm not always perfect. I still remember a day when a mentee walked into my office.

She said, "I'm so tired. It's such a hard day."

And I said, "I know. Mondays can be hard. I never sleep well on Sunday nights."

She stopped, startled, and then said, "I never realized that senior people have bad days too!"

Two Principles in One

In 2015, Jeff wanted to introduce a new Leadership Principle: Be Curious.

But he also wanted to keep the total number of LPs at 14, so one had to go. In the end, the casualty was Be Vocally Self-Critical. As you can see, it was folded into Earn Trust. The current version of the principle blends both ideas.

Earn Trust (Current Principle)

Leaders listen attentively, speak candidly, and treat others respectfully. They are vocally self-critical, even when doing so is awkward or embarrassing. Leaders do not believe their or their team's body odor smells of perfume. They benchmark themselves and their teams against the best.

Earn Trust of Others (Older Principle)

Leaders are sincerely open-minded, genuinely listen, and are willing to examine their strongest convictions with humility.

Be Vocally Self-Critical (Older Principle)

Leaders do not believe their or their team's body odor smells of perfume. Leaders come forward with problems or information, even when doing so is awkward or embarrassing. Leaders benchmark themselves and their teams against the best.

LP #12

DIVE DEEP

Leaders operate at all levels, stay connected to the details, audit frequently, and are skeptical when metrics and anecdote differ. No task is beneath them.

To: Amazon Employee
From: Jeff@amazon.com

?

—Jeff

The most dreaded email at Amazon is a single question mark from Jeff. These emails are Jeff's way of saying "WTF?!"—and they need to be answered immediately. Teams drop what they're working on to pull together a detailed investigation, usually within a few hours.

But these aren't about satisfying the curiosity of an imperial Jeff. They're about understanding how one of Amazon's processes failed and delivered a poor experience to a customer. This is exactly what the Dive Deep principle refers to when it says that anecdotes should not be ignored when they differ from metrics. Metrics give you the big picture. Anecdotes often show you where that picture is incomplete—or wrong.

Here's an example of a real "question mark" email:

To: Amazon Marketing Team

From: Jeff@amazon.com

Subject: FWD: Continue your search for "Sexual Lubricants"

?

--- Full email below ---[80]

What was happening here? In an effort to increase sales, Amazon's marketing team had been sending reminder emails to customers about unfinished purchases. If you were shopping for a birthday gift and got pulled away before checking out, a helpful nudge might be appreciated. The overall idea makes sense. This is the kind of **LP #8: Think Big** and **LP #9: Bias for Action** behavior that made Amazon great.

But Amazon shouldn't be sending emails about every abandoned cart. In this case, the customer had been browsing for sexual lubricants. When they received the embarrassing follow-up, they quickly forwarded it to Jeff. This isn't something anyone wants resurfacing in their inbox. While the overall process may work well most of the time, this particular email should never have been sent. Jeff's one-character response said it all: *We need to fix this—NOW.*

On the surface, it might seem like a minor slip. But Jeff saw something bigger underneath. Most companies send the occasional awkward email—it comes with the territory. But Jeff didn't want Amazon's emails to feel like everyone else's. He wanted every message to feel like it came from a trusted advisor. That kind of trust is hard to earn—and incredibly easy to lose.

I learned this while working in the credit card industry. You might have noticed that, even though almost everything else has gone digital,

[80] Brad Stone, *The Everything Store: Jeff Bezos and the Age of Amazon* (New York: Little, Brown and Company, Kindle Edition, 2013), 325.

we still get a lot of paper junk mail from banks and financial services companies. Why haven't they just switched to email like everyone else? Because it's actually *cheaper* to send a piece of paper.

Emailing someone about a credit card isn't free. You still have to do the targeting, selection, and compliance work. And over time, people have gotten better at spotting promotions—and more cautious about scams. Most of those emails go straight to spam. So companies try something else. They send you a piece of mail. It costs more, but that's the point. The expense signals that it matters, that it's worth your attention—and maybe even worth opening.

Now imagine you're Amazon, and you want to sell *everything*. From books to groceries to highly personal items. If customers can't trust you to use their data thoughtfully, they won't feel comfortable buying from you. Worse, they'll start thinking twice before opening the app. They'll shop around instead of heading straight to Amazon.

To be the customer's first stop, Amazon has to be the one they trust without thinking. That's why Jeff's "?" matters here. It's not about blame—it's about urgency. Something's off. Dive deep. Make it right.

The Five Whys

As we discussed in **LP #7: Insist on the Highest Standards**, the Andon cord is a tool used in manufacturing to stop the production line when there are quality problems. People often think of the Andon cord as a panic button. But it's actually a tool for early intervention—to flag a potential issue before it turns into a bigger one. Pulling the cord sends a signal to a team leader. The line continues for about 60 seconds while the team leader works to diagnose the problem. If the issue is resolved—as it is the vast majority of the time—the cord is pulled a second time, and the line keeps moving. It's a mindset built on curiosity—slow down, ask the right questions, and get to the root cause.[81]

---

[81] Amy C. Edmondson, *The Right Kind of Wrong.*

The Andon cord originates from Japan. After World War II, Japan was focused on rebuilding its manufacturing sector. The devastation of the war had left Japanese industry in ruins, and there was an urgent need to modernize and enhance production capabilities to rebuild the economy and compete globally.

To increase quality, Japan turned to W. Edwards Deming. Deming, an American statistician and quality control expert, introduced statistical process control methods to Japanese manufacturers. His contribution was so significant that he was awarded the Order of the Sacred Treasure by the Emperor of Japan. His teachings emphasized the importance of quality management, continuous improvement, and reducing variability in production processes.[82]

Japanese companies like Toyota embraced Deming's lessons and pushed them further, involving the entire organization in a culture of continuous improvement. These practices were eventually packaged as LEAN manufacturing and Total Quality Management, then brought back to the U.S., where they reshaped modern industry.

While Jeff has his own version of the Andon cord with his question mark emails, a more common tool for diving deep is the Five Whys. The idea is simple: keep asking "why" until you get past the surface and find the real issue underneath. It's how you make sure you're solving the right problem—the one that actually matters. The same approach can help uncover what customers truly need, leading to better decisions across everything from cost to experience.

It's a peculiar habit at Amazon, asking question after question like a persistent five-year-old. Not to be annoying—but because the first answer is almost never the whole story. Here's an example of how Jeff asked the Five Whys to investigate a safety issue:[83]

---

[82] "W. Edwards Deming," *Wikipedia: The Free Encyclopedia*, accessed April 1, 2025, https://en.wikipedia.org/wiki/W._Edwards_Deming.

[83] Peter Abilla, "Watch Jeff Bezos Use the 5 Whys Technique to Find Root Causes," *LinkedIn*, August 9, 2016, https://www.linkedin.com/pulse/use-5-whys-find-root-causes-peter-abilla/.

Jeff's Example of the Five Whys

The Situation: An associate had damaged his finger on a conveyor belt.

Why did the associate damage his thumb?

Because his thumb got caught in the conveyor.

Why did his thumb get caught in the conveyor?

Because he was chasing his bag, which was on a running conveyor belt.

Why did he chase his bag?

Because he placed his bag on the conveyor, but it then turned on unexpectedly.

Why was his bag on the conveyor?

Because he needed a place to put his bag, and there was no table nearby.

Conclusion: The associate injured his thumb because he didn't have a place to put his bag, so he used the conveyor as a makeshift table.

How to Fix It: Provide designated storage areas or signage directing associates to place maintenance bags on the floor.

It's easy to say Amazon should've had a storage space in place from the start—but that misses the point. The key isn't blaming someone for not getting everything right the first time. That's an impossible—and ultimately counterproductive—standard. The goal is to dive deep, understand what happened, and make it better.

You might've noticed the Five Whys example above only had four questions. That's fine. It doesn't have to be five. Sometimes it's four, sometimes six—occasionally more. Like most of these rules, it's not about following it exactly. Five Whys is about digging in until you get to the root cause.

Swimming on the Surface

Diving deep doesn't happen at most companies. People take a narrow view of their role. They are only responsible for what's in their domain. When problems arise, they want them to go away as soon as possible.

For example, I once worked at a company where a key client-facing system went down for 36 hours. Customers were unable to access their accounts, turning a technical failure into a major reputational blow. Our sales team carried the scarlet letter of that outage into every renewal conversation for the next year.

At the post-mortem, one of our senior tech leaders gave his update. "I have good news," he said. "It wasn't our fault. The outage came from a router change at corporate—they should've given us a heads-up."

I was stunned. Saying that it wasn't our fault completely missed the point. Just because we didn't cause the issue didn't mean that it wasn't our responsibility. Our customers didn't care where the problem started—they just knew the service was down.

So we started asking the Five Whys:

1. **What kicked off the problem?**
   A router change was made by the network team.
2. **Why did that change cause our system to fail?**
   The router change caused our system to rapidly spin up new processes.
3. **Why couldn't we recover quickly?**
   Those processes overwhelmed the system, causing it to run out of memory and start thrashing.
4. **Why didn't our fallback plans work?**
   We couldn't connect to the system to intervene.
5. **Why were we locked out entirely?**
   Once the system ran out of memory, it froze completely. And without a recovery layer in place, we had no way to get back in.

At that point, we had a problem we could actually solve. We didn't need to assign blame—we needed a better way to make sure this didn't happen again. So we added a lightweight management layer that would let us get back in and restart the system when it locked up.

Diving Deep at Work

I used this technique to solve a much bigger problem at work. It was my first role as a product manager, and I was excited. I had been assigned to one of the company's largest, most ambitious initiatives. The goal was to build a transformational online experience—one that would handle everything from low-level operational tasks to high-level executive reporting. It even included a personalized news feed for executives to reduce their need to jump between multiple tools. The vision was bold and energizing. It felt like we were building something that could redefine the industry.

Everyone was fired up about the project. But after a few months, something felt off. The enthusiasm was there, but progress seemed... fuzzy. So I started diving deep using the Five Whys.

1. **How's the project coming?**
    "Pretty well!" I was told. "Well... considering no one will give us enough money to do it right." That was my first red flag. So I kept digging.
2. **What's the problem with the budget?**
    "It's a big project, and we only got half of what we asked for this year. So you can't expect us to get everything done."
3. **How does this affect our timeline?**
    That's when things got quiet. Despite not receiving full funding, no one had adjusted the timeline to the new reality. I did a quick back-of-the-napkin calculation and realized we were moving at about a third of our original pace. That took

our 5-year plan and quietly stretched it into 15. Way too long for any stakeholder to accept—if they actually knew. Maybe no one else had run the numbers because they didn't want to see where they led.

4. **Why is this taking so long?**

That's when the real issue emerged. We weren't just building a new system—we were also upgrading our outdated tech stack. These two projects had been merged into one massive initiative, which ballooned our scope and slowed us to a crawl. It was like trying to drive a racecar down the track *while* rebuilding the engine.

But the truth was, we didn't have to do both at once. By separating them, we could limit the scope of the technology upgrade and dramatically reduce costs. Any improvements to the system could be added incrementally to the new platform.

5. **Is there another option?**

This produced two very different answers. From the technical team, it was a resounding "yes." They suggested we automate the platform transformation process—keeping the same features while moving from outdated Java Applet technology to a modern, HTML-native, service-oriented architecture. This proposal would streamline the transition, cut manual effort, and reduce costs. On the business side, there was plenty of skepticism. One colleague put it best when he said, "If it were really that simple, wouldn't we have done it already?"

We needed to prove the new approach could work. So we built a small prototype. The results were very impressive. The new approach cut costs by 80% and dramatically expanded the scope of what we could deliver. We could now do more with fewer resources.

The Result: That pilot was enough to convince the business team. It showed that our idea wasn't just technically viable—it was transformative. We'd found a path forward that worked, and the project turned around.

Diving Deep at Home

I've brought the principle of Dive Deep into my personal life. By slowing down and paying attention to the little things, I've unlocked a whole world hidden in the everyday. This became especially important during the COVID lockdown.

During lockdown, I developed a morning ritual. I'd get up early and go for a run, with one goal: take a photo of something inspiring.

At first, it felt strange—like I was forcing inspiration. Normally, I moved through the world without really seeing it. But when I started actively looking for beauty, I started finding it. The angle of light on a building. The color clash in a flowerbed. These small moments became daily doses of unexpected art.

Over time, I realized that beauty is everywhere—if I'm willing to look. When I walk through a museum and see Cezanne's *Basket of Apples*, I'm struck by how he saw those apples not just as fruit, but as color, form, light, and quiet presence. He didn't paint them to impress; he painted them to understand them. And I've found that if I look at a real bowl of apples in the right way—with attention, patience, and curiosity—I can see that same beauty. It's not about the subject. It's about how you see.

I used to think that beauty was all about perfection. Everything just so. Nothing out of place. But the more I pay attention to the world around me, the more I appreciate the beauty in what's a little off, a little unexpected.

One morning during a run, I stopped to take a photo of a flowerbed in someone's front yard. The colors were vibrant, the light hit just right—it felt like a little moment of art. Later, I showed the photo to a friend, who immediately said, "Oh no, there's a weed right in the middle. You've got to pull that out!" For a moment, I agreed. Maybe I should've cleaned it up. Made the world a little more perfect.

But then another friend looked at it and said, "I love it because of the weed. The imperfection makes everything else stand out. It feels more real."

Final Thoughts

Dive Deep is about getting to the root of things—whether you're untangling a problem at work or just trying to better understand the world around you. It means not settling for the easy answer. It's about staying close to the details, asking the harder questions, and doing the hard work to figure out what's really going on.

It's rarely convenient. It means slowing down when everything around you is trying to speed up. It means paying attention to things that don't seem like they matter—until they do. And it means questioning the answers everyone else is willing to accept at face value.

It can feel pedantic, or even a little bit silly. During lockdown, we'd left the city and were staying at my parents' house. Each morning, I meditated in their small backyard, focusing on the wind blowing through the leaves, the butterflies, the koi pond rippling in the sun.

I fancied myself a modern-day Henry David Thoreau. Just as he immersed himself in the natural world around Walden Pond, I immersed myself in the natural world of my parents' backyard. I took inspiration from his writings and tried to live more deliberately, finding joy in the simplicity of my surroundings.

But I also felt a little self-conscious. Like I was playing a part—breathing in the fresh air and pretending it was some profound act of wilderness living. I hadn't built a cabin. I wasn't off the grid. My parents were just inside making coffee. It all felt a little too comfortable, and I started to question how real any of it was.

Then I learned that Thoreau wasn't exactly roughing it either. His famous cabin? Built on land owned by his friend Ralph Waldo Emerson, who often had him over for dinner. Every Sunday, his mother and sister brought him a basket of bread and donuts. He took trips into town. He had guests. That great champion of self-reliance had room service.

At first, that felt like a letdown. But the more I thought about it, the more it made sense. It made him more human—and more like me. That's the thing about diving deep. You trade the clean version of the story for the real one. It's more complicated, but often better.

LP #13

HAVE BACKBONE; DISAGREE AND COMMIT

Leaders are obligated to respectfully challenge decisions when they disagree, even when doing so is uncomfortable or exhausting. Leaders have conviction and are tenacious. They do not compromise for the sake of social cohesion. Once a decision is determined, they commit wholly.

If you have conviction on a particular direction even though there's no consensus, it's helpful to say, "Look, I know we disagree on this, but will you gamble with me on it? Disagree and commit?"

—Jeff, 2016 Amazon Letter to Shareholders

It's 2004, a year after Steve Jobs invited Jeff to Cupertino to show him the iPod prototype. Jeff responded by creating a digital media business led by Steve Kessel.[84] But what type of digital media business was it going to be? Amazon was still known as the world's largest online bookstore. Now it was moving into digital media. There was no roadmap for this, and the debates on how to move forward were

---

[84] See LP #2: Ownership.

intense. Should Amazon focus on digital books, music, or video? Should it build its own hardware or partner with existing manufacturers?

Kessel encouraged open dialogue and dissent, ensuring all voices were heard. Some were skeptical about Amazon's ability to succeed in the hardware market, given its e-commerce roots. Bill Carr, an early member of the team, couldn't fathom building a hardware device. He told Kessel, "We're an e-commerce company, not a hardware company!" He thought, "We should partner with third-party equipment companies that were good at designing and building hardware, and stick to what we knew: e-commerce. I regularly pointed out that he knew nothing about hardware—he wasn't a gadget guy, and his ancient Volvo didn't even have a car stereo."[85]

That was when the project was still in the "Have Backbone" phase where dissent is appreciated and even encouraged. After listening to all of these viewpoints, Kessel, as the leader, decided that Amazon needed to either build this capability or buy another company that could do it. Since it was core to Amazon's future, it couldn't be outsourced. Now the team was in the "Disagree and Commit" phase. Once the decision was made to build the device, Carr and the rest of the team committed to the new plan.

In the "Disagree and Commit" phase, everyone had to commit their heart and soul to the project. When Kindle was released, it met its goal. Users could download eBooks from anywhere, and the screen looked far closer to paper than a typical digital display. But the job wasn't done. The first-generation Kindle was big and clunky, mainly because it required an old-school cellular radio to download books. With each version of the Kindle, the team went through the same process of "Have Backbone; Disagree and Commit," incrementally improving the product. Today, the Kindle—and the Kindle ecosystem—has completely changed the way people read.

---

[85] Colin Bryar and Bill Carr, *Working Backwards*, p. 178.

Digging Into the Principle

This is probably the most misunderstood of Amazon's Leadership Principles. At its heart, it's a simple framework we all understand: separating planning from execution and engaging the team differently at each phase. During the "Have Backbone" phase, it's valuable to encourage disagreement to improve the decision. But once the team enters execution, everyone must "Disagree and Commit" and work together. Team members need to be humble, putting their egos aside for the good of the team.

While we intuitively know this approach is right, it's hard to practice. Often, companies (and people) refuse to accept reality and just drive a project forward without thinking. Leaders may look at a project and say, "This needs to happen!" They're practicing a form of magical thinking—imagining that if the risks aren't acknowledged, they won't occur. I remember a friend capturing the mood perfectly (and satirically) during a slow-moving disaster of a project. With quiet resignation, he said, "We've explored all the workable options and found none. So now we're doubling down on hope and prayer." Let's look at an example of how magical thinking can sink a project.

Senior management was in a bind. They'd already committed to a platform upgrade that had spun out of control—cost overruns, scope creep, poor planning. Still, they held out hope. On paper, it still looked like a winner. They just needed a way to finish it. They were desperately searching for a "get out of jail free" card.

And they found one—or so it seemed. The idea was to take their internal platform, package it as a commercial product, and sell it to other companies. Even better, there was already an interested customer. The deal looked great: a large upfront payment and an ongoing maintenance fee. This would help offset the shortfall from the original project. It felt like the perfect fix. They said, "How could it go wrong?" with the confidence of people who'd already decided not to look too closely.

Once the contract was signed, the company threw everything into delivering for that one customer. It was expensive—far more than they ever earned from the deal. To meet the deadlines, they redirected the entire division's tech budget to support this single client for a full year. All development shifted to the client's specific needs, and while they built a fantastic system, it wasn't the one they needed. The original platform—the upgrade they'd set out to complete—slipped even further behind.

Amazon takes a radically different approach. It flips this dynamic on its head. Instead of committing to a project prematurely and then facing disagreement during execution, it encourages all disagreement up front. Then, once a decision is made, every team member commits fully.

Here's how it works:

- **Have Backbone:** When making a decision, each team member is obligated to speak up with any objections they have. This can feel uncomfortable at first, because it is actually courting conflict in the team. But when I first saw this in action, I found it refreshing. Instead of the "get along at all costs" culture found in many companies, this approach was far more outcome-focused. It's a healthy kind of conflict—not between people, but between ideas. It's about voicing concerns, offering alternative viewpoints, and challenging assumptions. By stress-testing every idea, the team ensures that the best decision is made.

- **Disagree and Commit:** After the decision is made, everyone comes together, regardless of their earlier opinions. Some people won't get their way. The process isn't about compromise and pleasing everyone. It's about setting a single opinionated direction and moving forward together. Each team member commits to the plan as if they had supported it from the beginning. The team becomes one collective organism, each person doing their part to support the larger mission.

Once you understand how this principle works, you start noticing it in all kinds of places. Say you're planning a trip with friends. Everyone has a preference for where to go. Some want to go to the beach and others may want to go hiking—and the debates can get pretty heated. But once the destination is chosen and the tickets are booked, that's it. You may not have gotten your first choice, but it's time to just enjoy it. We also see it at work when a boss leaves. Multiple people vie for the job, each with their own vision and style, and things can get pretty intense. But once someone gets the role, the rest of the team gets behind them and moves forward with their new jobs.

Have Backbone

There's an old saying. If a ship sets out on a thousand-mile journey and it steers true, it will arrive at its destination. But if the navigation is off by as little as one degree, the ship will miss its mark by miles. The same principle applies to projects. While the most important decisions are the ones made at the very beginning, it's easy to skip the early steps and jump right into execution. Teams put plans together and start working as quickly as possible. Little issues start to creep up, but everyone tells the boss, "Things are going great. Don't worry, we'll be able to fix things later." But as the problems get bigger and the project gets closer to completion, the cracks get bigger, deadlines get pushed, and budgets are overrun.

But it doesn't have to be this way. Instead of waiting until the end of the project for these problems to fester, we can pull them to the front. This is difficult and requires people to act differently. It's natural for teams to want to avoid conflict, especially early on when everything seems to be going smoothly. However, this is precisely the time when disagreement is most valuable. To make the most of this process, the team needs to shift its mindset from focusing on who's right to focusing on making the best decision for the group. This means moving away from seeing decisions as personal battles and instead valuing the strength of diverse perspectives.

Encouraging team members to take a stand, even on less popular sides of an issue, forces everyone to think more critically and consider alternative viewpoints, ultimately leading to better decisions. At Amazon, disagreements are encouraged as a way to challenge ideas, not people. When someone disagrees passionately, it's seen not as a disruption but as a sign of commitment to improving the idea. This changes the way that the team handles conflicts, treating them as opportunities rather than obstacles.

Amazon has mechanisms to channel disagreement and keep the focus on the idea rather than the person. That's one of the key reasons Jeff banned PowerPoint and moved to the PRFAQ process. This happens when people draft PRFAQs (see **LP #1: Customer Obsession**) during the Document Review process. Rather than trying to win people over to their side with a PowerPoint presentation, people write documents that are reviewed by a group so everyone can improve them. Even Andy Jassy. When Andy came up with the idea for Amazon Web Services, he went through 31 revisions of his document.

I've started doing this during debates at work. When we break into two sides to discuss an issue, one side is usually the clear favorite. When everyone joins that group, there's less to challenge, and the conversation tends to be less engaging. So I've begun choosing the underdog position more often, saying, "Maybe it's worth exploring this a little further—just to make sure we're not missing anything." It's harder, but it pushes me to think more creatively. I have to make arguments others haven't considered and, in the process, help the majority side stretch their thinking a little further too.

Disagree and Commit

Now we move to the next phase: Disagree and Commit. It's not about stubbornly sticking to a bad idea. It's about executing something really well and doing everything possible to make it successful. Jeff gave the following example in Amazon's 2016 Annual Letter:

> We recently greenlit a particular Amazon Studios original. I told the team my view: debatable whether it would be interesting enough, complicated to produce, the business terms aren't that good, and we have lots of other opportunities. They had a completely different opinion and wanted to go ahead. I wrote back right away with "I disagree and commit and hope it becomes the most watched thing we've ever made."

It's not about compromise. You can't make a good movie by listening to everyone's point of view. If you try, you end up with something like the movie *Cats*—a film that tried to be a faithful stage adaptation, a mainstream crowd-pleaser, and a digital effects showcase, and instead gave us human-cat hybrids that made everyone uncomfortable. Everyone who gave notes is happy, but no one wants to see the movie. That's not how you make a movie. You choose a clear direction, and hundreds of people work together to bring that specific vision to life.

Disagree and Commit helps prevent teams from falling into the trap of overanalyzing a project. Sometimes it feels like we're caught in the paradox of Buridan's Ass, named after the French philosopher Jean Buridan. In this story, a donkey is placed exactly midway between two identical piles of hay. Unable to decide between the two equally appealing options, and with no compelling reason to choose one over the other, the donkey eventually starves to death. That's how I feel at work when I'm stuck overanalyzing a project—paralyzed by indecision, endlessly weighing options, and getting nowhere. Disagree and Commit cuts through that paralysis, allowing the team to make a decision, move forward, and focus its energy on making it work.

But it's not just about trying new things; it's about deeply understanding what works and what doesn't. To truly test ideas, the team must be fully committed. Without complete dedication, it's impossible to determine whether a plan failed because the concept was flawed or because the effort wasn't sufficient. That's why wholehearted support for a project is crucial—it's necessary to genuinely assess its viability. By committing fully, teams can effectively test ideas, learn from failure, and ultimately achieve better outcomes.

When you commit to an idea, you give yourself the opportunity to test it. You create the chance to be wrong—and in doing so, to learn something valuable. I experienced this recently at the Tower of London. A teenager in our group, using their knowledge of New York fashion, confidently asserted that the people in front of us were from New York. They based this on the fact that the group was wearing North Face jackets and black-and-white Nike "Pandas." It seemed reasonable, but I was curious to test it.

As we got closer, expecting New York accents, we instead heard them speaking French. It turned out that this stylish group was from Paris. By forming a hypothesis and testing it, we learned that people from Paris can look and act a lot like people from New York. This simple experiment reinforced the idea that committing to an assumption and testing it—whether in fashion or in projects—can lead to surprising insights and deeper understanding.

A Rough and Tumble Sports Team

If this seems hard and demanding, that's because it is. Jeff is famous for loving hard work. He originally wanted to name the company "Relentless" and still owns the domain name.[86] Remember **LP #9: Frugality**—when we talked about how Amazon doesn't pay for business class to India? It's like that. It's a respectful culture and gets the job done quickly and efficiently, but it's not "nice." This Leadership Principle says a lot about Amazon's focus. The company prioritizes finding the right answer and executing it effectively, even if it means engaging

---

[86] Fun Fact: If you go to relentless.com, you are redirected to Amazon.

in "uncomfortable or exhausting" discussions. Amazon clearly values being right over being nice, as reflected in the statement, "They do not compromise for the sake of social cohesion." The cohesion comes afterward—once a decision is made, everyone commits fully to it.

Employees often see their companies like a family. However, Amazon and other tech companies take a different approach. Netflix captures this mindset well. Netflix CEO Reed Hastings and Chief Talent Officer Patti McCord spent years shaping the company's culture, which they famously outlined in a PowerPoint presentation that went viral, amassing over 100 million views. One slide says, "We're not a family but a team. We're like a pro sports team, not a kids' rec team."

This mindset was clear in 2012 when McCord, who had helped build the company with Hastings for 15 years, was let go as Netflix became a media company rather than a DVD shipping company. Despite everything she had contributed, it was no longer a fit for her talents and expertise.

She later said it felt like a breakup—painful, personal, and hard to let go of. But it also reflected the culture she had helped create. Netflix wasn't a family; it was a team built to adapt and win, even when that meant making tough, unsentimental choices.

Final Thoughts

Like most good ideas, Have Backbone; Disagree and Commit sounds easy when you first hear it. It's harder in practice. First, you bring everything you have to the decision-making process—your doubts, your insights, your willingness to challenge the easy answer. That's what Amazon did with the Kindle, working through hard questions about hardware, capabilities, and strategy. Once the decision was made, the team didn't linger on disagreements. They moved forward, fully committed, improving the product step by step until it reshaped an entire industry.

We've also seen what happens when this principle gets ignored. In a lot of companies, decisions get rushed through on optimism and wishful thinking, only for disagreements to bubble up later—during execution, when it's already too late. Resources get wasted. Projects drift off course. The original vision gets buried under a pile of quick fixes and compromises.

Amazon's approach flips that cycle. It encourages real debate early and demands full commitment later. Teams face the hard truths when they're still manageable and build momentum when it matters most.

Using this process outside of Amazon can be tricky. The company takes this principle to the extreme, placing the highest importance on finding the right answer and executing it efficiently, even if that means engaging in "uncomfortable or exhausting" discussions. For me, though, pushing for the right answer at the expense of social harmony usually isn't the best approach.

When I apply this principle, I tend to use it in a more subtle way. It's about separating the planning phase from the execution phase, making it clear that we should act differently in each one. It helps ensure decisions are made thoughtfully, and then carried out without constant second-guessing or changes in direction.

At work, we follow an iterative development process, where feedback and adaptation are ongoing. While it might seem like we're always planning and executing at the same time, there are subtleties to it. We break down the work into two-week increments called sprints. During sprint planning, we encourage everyone to share differing opinions and discuss different approaches. This open dialogue helps us find the best path forward. Once we move into execution, the decisions have already been made, and we can proceed smoothly. Then, two weeks later, we repeat the process.

In my personal life, I've found good opportunities to apply Have Backbone; Disagree and Commit. Sometimes it's important to Have Backbone—like when you're assembling furniture and realize that two brains are better than one when it comes to reading the instructions. Speaking up early can save a lot of time—and a few missing bolts—later. It's like that ever-present IKEA cartoon in the instructions that quietly reminds you to always bring a friend.

But just as often, it's about knowing when to Disagree and Commit. In many situations, there's no single "right" way to do things. Think about the classic argument couples have over how to load the dishwasher. Each partner has their own method, but at some point, one person just has to Disagree and Commit so the dishes get done and peace is kept. You can fight the dishwasher battle, or you can stay married—but probably not both.

LP #14

DELIVER RESULTS

Leaders focus on the key inputs for their business and deliver them with the right quality and in a timely fashion. Despite setbacks, they rise to the occasion and never settle.

"We are stubborn on vision. We are flexible on details."

—Jeff

To me, Deliver Results belongs at the end of the Leadership Principles. When I was at Amazon, it *was* at the end—there were 14 principles then, starting with **LP #1: Customer Obsession** and ending with Deliver Results. Everything in between was about how you got from one to the other. So I'm going to treat it that way in this chapter.

We begin with **LP #1: Customer Obsession**—relentlessly understanding the customer we're targeting and the problem we're trying to solve. From there, the next 12 principles are the tools that help us to drive results:

- Break down the problem into pieces that can be owned by individual teams. **(LP #2: Ownership)**

- Figure out what you need to build—whether it's a process or a product—and find the simplest way to build it. **(LP #3: Invent and Simplify)**
- Trust your gut but always strive to improve. Admit when you're wrong. **(LP #4: Are Right, A Lot)**
- Don't be afraid to explore new territory. **(LP #5: Learn and Be Curious)**
- Protect your culture. Don't let short-term pressure compromise long-term success. **(LP #6: Hire and Develop the Best)**
- Deliver on your commitments to the highest possible standard. **(LP #7: Insist on the Highest Standards)**
- If you want to do something big, you'll need to experiment— and that means failing a few times first. **(LP #8: Think Big)**
- Failure and course correction are expected. Speed matters more than perfection. **(LP #9: Bias for Action)**
- You'll never feel fully finished. Prioritize your resources anyway. **(LP #10: Frugality)**
- Your brand is built by acting consistently. That defines what others expect from you. **(LP #11: Earn Trust)**
- Dig deep to understand what's really going on, not just what's on the surface. **(LP #12: Dive Deep)**
- Have honest, difficult conversations early. Once a decision is made, commit and move forward together. **(LP #13: Have Backbone; Disagree and Commit)**

At Amazon, these are all tools to help you Deliver Results. The ultimate goal is always customer impact—and while process matters, it should never become the focus in and of itself.

Success is rarely straightforward. It takes effort, and there aren't any shortcuts. It's not like some flash of brilliance suddenly hits and now you've got a Kindle. Delivering results means embracing the process: breaking down challenges, experimenting with solutions, learning from setbacks, and continuously refining along the way. It's a path of persistence and hard work that, over time, can lead to some great outcomes.

Amazon's Brute Force Algorithm

If delivering results for customers is what truly matters, why do we need the other 12 principles? Why not just focus on the customer and get things done? We need a framework to get from one to the other. Trying to deliver results without a framework or a plan is like building a house without a blueprint.

Amazon's approach to delivering results has been described as a "brute force algorithm." It's a good description, borrowed from computing, that means trying every possible solution until something works. At first glance, this might not sound like Amazon—a company known for innovation and efficiency. But if you think about Jeff's quote, "We are stubborn on vision. We are flexible on details," it starts to make more sense.

It's easy to focus on the first half of that statement—being "stubborn on vision" feels bold, almost heroic. It resonates with anyone striving toward a big goal. But the second part—being "flexible on details"—is where the real work happens. It's not just about having the idea. Most of the effort lies in execution. Amazon relentlessly explores every possible road to get there. It's a process of trial and error, with many failures along the way, that leads to real breakthroughs.

Take Amazon's long-running effort to profitably sell everyday essentials—the kinds of items you'd find in a drugstore: milk, batteries, diapers, and cleaning supplies. These are things people need quickly, but they come with razor-thin margins, and fast delivery often costs more than the products themselves earn.

Over the years, Amazon has tried multiple strategies to crack this market, including:

- **Amazon Fresh (2007):** A grocery delivery service

- **Subscribe & Save (2007):** A subscription model offering recurring essentials at a discount
- **Prime Pantry (2014):** A service bundling bulky household items into a single box for a flat fee (discontinued)
- **Prime Now (2014):** An ultra-fast delivery service promising one- or two-hour delivery in major cities (discontinued)
- **Whole Foods Market (2017):** Acquired for $13.7 billion to expand into physical grocery retail
- **Amazon Go (2018):** Cashier-less convenience stores using "Just Walk Out" technology

And that's just the visible part. Behind the scenes, Amazon has poured billions into its logistics and delivery infrastructure to make shipping faster and more reliable. With many Prime items now arriving in under a day, it seems they've finally cracked the code on everyday essentials.

Ben Thompson, writing in his blog *Stratechery*, explains why he now buys his basics from Amazon:

> "Yes, I could go to the store *right now* and get what I need, but (1) I'm probably not going to go right now, and (2) I might forget or be pressed for time later; however, if I order from Amazon *right now* I know I will have what I need either by the end of the day or when I wake up in the morning. It feels silly to even consider going to the store unless I need the item in question in the next hour."[87]

Amazon didn't crack the code on everyday essentials through sheer force of will. It wasn't about having the perfect strategy from the start. It was about testing, adjusting, and staying flexible enough to pivot when something wasn't working. Progress almost never happens in a straight line. It happens by continually driving forward, making corrections, and being willing to adapt to the reality in front of you.

---

[87] Ben Thompson, "Google Decision Follow-Up, Amazon Earnings," *Stratechery*, August 7, 2024, https://stratechery.com/2024/google-decision-follow-up-amazon-earnings.

Don't Fight the World; Embrace It

Amazon's success comes from working with the world, not against it. That mindset—adapting to reality rather than resisting it—has shaped the company from the beginning. But it's not just a business principle. It's something I've seen play out again and again in life.

You can't force things to go your way—no matter how badly you want to. People used to come to us at Amazon and say, "I really want to be as innovative as Amazon," but they weren't willing to do the hard work. The world doesn't care about your plans, your effort, or how much you want something to happen. Whether it's business or life, you have to work with what's real. Fighting reality doesn't get you anywhere. Adapting to it is the only way forward.

I learned that lesson in an unexpected place: on a ski mountain.

Skiing teaches you quickly that you can't fight the mountain. The conditions are always changing—sometimes it's ice, sometimes powder, sometimes both. If you try to impose your will on the terrain, you're going to fall. The trick is to lean into what's there, adapt, and trust the process. Skiing isn't about conquering the mountain—it's about working with it.

I wasn't always that way. When I first started skiing, I thought it was all about conquering the mountain. My goal was to dominate those expert runs and prove my superiority. My 14-year-old son is in that phase now—he's all about the double blacks and complains when the conditions aren't perfect. But that's not what skiing is about. It's about taking what's in front of you and finding your rhythm, trusting your skis, and flowing with the mountain instead of fighting it.

Over time, I've learned to change my mindset. Now, instead of chasing the hardest runs or the steepest slopes, I focus on the quality of each turn. I don't worry if the terrain is icy or rough—I adapt. Whether I'm side-slipping down a chute, finding a smooth line through moguls, or gliding down a powder-filled cruiser, I let the mountain guide me. And when I do, I ski better than I ever have—and I enjoy it more, too.

Amazon has that same flexibility. Conditions change. Competitors show up. Markets shift. You can't fight what's happening. Instead, you adapt. You iterate. You experiment. You lean into the process and let it take you where you need to go.

This is where Amazon stands apart. Many see flexibility as a weakness—as indecisive flip-flopping. But that kind of inflexibility is just another word for stubbornness, and stubbornness doesn't win. The market doesn't care how much effort you put into the wrong idea. Amazon knows that flexibility is a strength. It's what lets you navigate the inevitable twists and turns and still come out ahead, whether you're tackling a business challenge, skiing down a mountain, or just dealing with everyday life.

Getting the Basics Right

Getting the basics right isn't exciting, but it's what really matters. It's easy to focus on the big, flashy stuff—solving tough problems, coming up with brilliant ideas—but at the end of the day, it's the fundamentals that make the difference. When you skip them, things fall apart. And the crazy part? Most people don't even realize they're skipping them until it's too late.

I learned this in high school. I was good at math. Actually, in the words of Mike from Dollar Shave Club, I wasn't good at math, I was f***ing great.[88] I was on the math team. I scored seventh in the Nassau County math championships in middle school—in a county bigger than the state of Rhode Island.

Then there was my friend Richard. Richard was good at math, but not as good as I was. He wasn't on the math team. But he would consistently outscore me in AP Calculus. I'd get something like a 95. I knew the material but made stupid mistakes. He'd beat me with a 99 or 100. Getting a perfect score on a calculus test isn't about who can

---

[88] Watch Mike talk about how f***ing great his blades are: https://www.youtube.com/watch?v=ZUG9qYTJMsI

solve the hardest problems—it's about who can solve the problems on the test with the fewest errors.

Richard became a doctor. He's a very good doctor. He's rock solid on the basics. While you'd think this would be true of all doctors, it's not. To become a licensed physician in the U.S., all doctors take the United States Medical Licensing Examination. You'd think they'd need a 90% or so to pass. Surprisingly, the passing grade is about 70%. And that's for people fresh out of med school—when they're the most up to date.

Not all doctors are as good as Richard. A few years ago, I had a palsy on the right side of my face. Richard lives in California, so I went to a local neurologist. I waited in his dank waiting room for 45 minutes, flipping through five-year-old magazines. Eventually, I was called into his office, where a younger doctor was shadowing him. The neurologist gave me a long speech: "In my experience, this type of palsy is often caused by a virus. So I'm prescribing you two weeks of antiviral medication."

When I told Richard, he said, "Man, this guy should keep up on his reading. I just studied to renew my boards and learned that the latest guidance is not to prescribe antivirals because we don't think the palsy is caused by a virus anymore." He pointed me to a few lines from the American Academy of Neurology (AAN), which said:[89]

- **Steroids (Strong Evidence).** For patients with new-onset Bell's palsy, oral steroids should be offered to increase the probability of recovery of facial nerve function.
- **Antivirals (Weak Evidence).** Patients offered antivirals should be counseled that a benefit from antivirals has not been established, and, if there is a benefit, it is likely that it is modest at best.

[89] American Academy of Neurology, *Practice Guideline Update Summary: Acute Treatment of Bell's Palsy*, last modified November 7, 2020, https://www.aan.com/Guidelines/home/GuidelineDetail/573.

With those two lines, the curtain was pulled back—like finding out that McDonald's secret Big Mac sauce is just Thousand Island dressing. All that confidence, all that experience, and it came down to one mistake: he hadn't kept up.

In medicine, in business, in anything that requires skill, the basics aren't something you master once. They're something you maintain. Richard knew that. The neurologist didn't.

Making It Work Every Day

Understanding this is one thing—putting it into action every day is another.

It's easy to have insights in the moment, but without a way to track them and follow through, they fade away. Delivering results in my own life means having a system that keeps me focused, honest, and moving forward. For me, that system is my Bullet Journal.

At first glance, it looks like an ordinary planner. But for me, it's something more. It's a tool that helps me bring meaning into the things I do each day. It lets me look back at the decisions I've made, look ahead at the ones coming, and ask myself a simple question: Am I living a good life—one with intention and meaning?

We make the same decisions over and over again, often without realizing it. We assume today's struggles are new, when in reality, they're usually just echoes of old choices. But when you can see the pattern, you can break it. A Bullet Journal helps me see those patterns clearly enough to do something about them.

At its core, it's not about documenting every thought or keeping a perfect record. It's about creating a throughline—a sense that today connects to yesterday and tomorrow. It helps me be more deliberate in how I live, intentionally living my life and making sure it matters.

It's a simple process, but it's powerful. Each day, I write down what I plan to do, what actually happened, and things I notice along the way. I jot down small moments—things that worked, things that didn't.

Over time, it creates a running story of my choices, my priorities, and my patterns. I'm not trying to capture everything. I'm trying to catch the things that matter—important events I want to remember, or key lessons that will help me live my life better. It keeps me honest about how I spend my time, gives me perspective, and keeps me connected to what matters most.

I supplement my Bullet Journal with another everyday tool. Each night, I ask myself a set of questions. It's not about what I accomplished or whether I hit some benchmark of success. It's more basic than that. I just ask, "Did I do my best today?"

The idea comes from Marshall Goldsmith, who calls it the Daily Questions method. Each day, I check in with myself to see if I'm living the way I say I want to live. Did I show appreciation for my kids? Did I think of two things I'm grateful for? Did I exercise? It's a straightforward practice, but it pulls the important things into focus.

Goldsmith took it even further. He hired someone to call him every night and ask him his questions out loud. I use an app called Way of Life to track mine. It's not complicated. But it creates just enough structure to make sure I don't lose sight of what matters.

Together, these two tools—the Bullet Journal and the Daily Questions—help me stay anchored. They keep me from drifting. They remind me that meaning isn't something I stumble into. It's something I build, day by day.

Final Takeaways

Delivering results isn't about finding a perfect plan and executing it without mistakes. It's about starting with a clear purpose, adapting along the way, and sticking with it when things get hard. At Amazon, the Leadership Principles give structure to that process, but at the heart of it is something simpler: focus on what matters, work with the reality in front of you, and keep moving forward.

The same is true outside of work. Whether it's building a career, raising a family, or just trying to live a meaningful life, results don't happen by accident. They happen by getting the basics right, learning from setbacks, and being honest about whether your actions match your intentions. There's no magic shortcut. There's only the work of showing up, adjusting as you go, and staying committed to building something that matters.

One of the clearest places where this shows up is retirement. You can't just show up one day and expect everything to fall into place. It takes planning, consistency, and a willingness to adapt when things don't go the way you expected.

When most people think about retirement, they generally focus on money. Once they've saved enough, they figure they're set. Retirement is seen as the finish line—finally, a break from work. *I'm free*, they think. But that kind of freedom isn't always what it's cracked up to be. As Sartre might say, they are "condemned to freedom." Having the ability to do whatever you want, whenever you want, might sound ideal, but in reality, it can feel more like a burden than a gift. At first, waking up with no schedule, no deadlines, and no obligations sounds perfect. But after a few weeks, that wide-open space can feel less like freedom and more like a void.

For some people (myself included), retirement without purpose feels a lot like unemployment. It's not just about having enough money—it's about having something meaningful to do.

Having a plan—or at least a clear goal—makes all the difference. My friend Michelle figured that out early. She realized that a fulfilling retirement boils down to three things: financial security, good friends, and making an impact. And she didn't wait until the last minute—she'd been working on her plan for over a decade.

Breaking it down, she had:

1. **Financial** – She invested in real estate that would provide a steady income for the rest of her life.

2. **Friends** – She joined a knitting group and a golf group, building a community beyond work.

3. **Impact** – Her son started a nonprofit focused on improving literacy for underprivileged kids. Michelle is using her business skills to help get it off the ground—and watching her grandkids while her son goes out to save the world.

When I saw everything she had put in place, I asked her, "How did you do it?"

She shrugged and said, "It's not that hard. I set a goal and figured out how to execute. I broke it down, built a plan, and made steady progress until I got there. This is the way I've learned to get things done, having spent my entire career as a project manager. Retirement was just another project."

LP #15 & LP #16

THE FINAL TWO PRINCIPLES

LP #15: Strive to Be Earth's Best Employer

Leaders work every day to create a safer, more productive, higher performing, more diverse, and more just work environment. They lead with empathy, have fun at work, and make it easy for others to have fun. Leaders ask themselves: Are my fellow employees growing? Are they empowered? Are they ready for what's next? Leaders have a vision for and commitment to their employees' personal success, whether that be at Amazon or elsewhere.

LP #16: Success and Scale Bring Broad Responsibility

We started in a garage, but we're not there anymore. We are big, we impact the world, and we are far from perfect. We must be humble and thoughtful about even the secondary effects of our actions. Our local communities, planet, and future generations need us to be better every day. We must begin each day with a determination to make better, do better, and be better for our customers, our employees, our partners, and the world at large. And we must end every day knowing we can do even more tomorrow. Leaders create more than they consume and always leave things better than how they found them.

When Jeff handed over the CEO reins to Andy Jassy in 2021, it was about more than a leadership change. Amazon was no longer in the image of Jeff. The Leadership Principles were always an extension of Jeff's personality and a way to instill his values in the rest of the company. Now that he was leaving the company, the Leadership Principles changed as well. Two new principles were added: Strive to be Earth's Best Employer and Success and Scale Bring Broad Responsibility.

Amazon has always been a reflection of Jeff's leadership style—intense, data-driven, and relentlessly focused on the customer. The original principles were about delivering for the customer at all costs. They encouraged employees to push themselves harder, think bigger, and constantly raise the bar.

For years, Amazon held firm at 14 Leadership Principles. Jeff wanted to avoid overwhelming employees—if a new principle was added, an old one had to go. So when **LP #5: Learn and Be Curious** was introduced, it quietly replaced the LP of Be Vocally Self-Critical. Be Vocally Self-Critical was then merged into **LP #11: Earn Trust**.

But most importantly, these new principles are fundamentally different from the original 14. The old Leadership Principles had a voice that was unmistakably Amazon—quirky, direct, and bold. These new ones? They feel like something from a corporate handbook. You won't find anything like "Leaders do not believe their or their team's body odor smells of perfume" here. Instead, they read like an effort to soften Amazon's edges—to come across as more responsible and a little less peculiar.

The first of the new Leadership Principles, Earth's Best Employer, signals this shift directly. It's a recognition that Amazon's culture didn't put employees first—and that needs to change. A more conventional Amazon could mean employees who feel valued, communities that see the company as a positive force, and a reputation that extends beyond being the fastest, cheapest, and most convenient option. When I was at Amazon, employees were unquestionably second-class citizens after customers. As I mentioned in **LP #10: Frugality**, there were no gourmet free lunches—perks that other tech giants take for granted. But it's not about the perks. Earth's Best Employer suggests something bigger: Amazon is committed to treating its employees with respect. That might not be peculiar, but it's progress.

The second new principle, Success and Scale Bring Broad Responsibility, marks a shift in Amazon's place in the world. It's no longer just a scrappy disruptor—it's a global giant with obligations that extend beyond customers and shareholders. For years, Amazon focused on anticipating customer needs and outpacing competitors. Now, it's being asked to also consider its impact on the environment, governments, and communities. This principle is an acceptance of a new reality: Amazon isn't just an innovator anymore; it's one of the world's biggest companies, and companies of that scale are held to different standards.

A Kinder, Gentler Amazon

The company has changed a lot since 1994, when Jeff was driving across the country with nothing but a dream of selling books on the internet. The Jeff of the mid-'90s isn't the Jeff of today. In 1997, he wrote in the first shareholders letter:

> It's not easy to work here. When I interview people I tell them, "You can work long, hard, or smart, but at Amazon.com you can't choose two out of three."
>
> —Jeff, 1997 Amazon Letter to Shareholders

But in 2018, he sounded very different. I guess that's what a hundred billion dollars will do to you:

> I prioritize sleep unless I'm traveling in different time zones. Sometimes getting eight hours is impossible, but I am very fo-

cused on it, and I need eight hours. I think better... When Amazon was a hundred people, it was a different story, but Amazon's not a start-up company, and all of our senior executives operate the same way I do.

—Jeff, Economic Club of Washington interview, 2018

Amazon's challenge is holding on to the creativity and innovation that built the company while adapting to the reality of being one of the world's largest companies. Speed, invention, and an obsessive focus on customers got it here—but scale changes things. A company this big can't operate like a startup forever. Regulation, scrutiny, and employee expectations demand a different kind of leadership.

This isn't easy for Amazon, a company that prides itself on having a startup mentality. Jeff always preached the importance of Day 1—that state of speed, innovation, and an obsessive focus on customers. Day 2, he warned, was what happened when a company got comfortable and stopped taking risks—when protecting what it had built became more important than building what came next. For years, Amazon fought against that fate, constantly reinventing itself to stay ahead.

As Amazon grows, it needs to be a different kind of leader—one that doesn't just chase innovation but also leads the world in responsibility. It's not about Day 2 or slowing down. It's about growing up.

When Technology Grows Up

As Uncle Ben tells Spider-Man, "With great power comes great responsibility." Small companies get the freedom to try new things—regulators look the other way, giving them space to move fast and break things. But as they grow, that leeway disappears.

I remember a conversation about this back in 2008, when social media was still in its awkward phase. Facebook was overtaking Myspace, but for most people, it was just a place to share pictures and play online Scrabble. A startup thought leader told me, "The U.S. government doesn't care about money laundering on social networks. I showed a regulator how to do it, and they didn't even care."

Technically, he was right—but only because the stakes were still low. Back then, most online commerce was happening on niche platforms like Second Life, and regulators had bigger things to worry about. But when internet-fueled crime became a real problem, the government took notice. By 2013, the FBI had arrested Ross Ulbricht, better known as Dread Pirate Roberts, for running the Silk Road—an online black market trafficking in drugs and other illegal goods. He didn't just get a slap on the wrist. He got two life sentences and a $183 million fine.[90]

When new technologies start out, they feel like novelties—exciting, experimental, almost like toys. A century ago, electricity and automobiles weren't just tools; they were futuristic wonders. But as they grew, they stopped being curiosities and became part of everyday life. Amazon followed the same path. Once a scrappy online bookstore, it's now infrastructure. And being essential means being accountable.

At a certain point, innovation stops being just about making things better and starts forcing uncomfortable conversations. That's exactly what happened in the early 1960s, when a medical breakthrough raised a question no one wanted to answer: If you can save a life—but not all of them—who gets to decide who lives? This wasn't just a technology problem; it was an ethical one.

A revolutionary new technology had just been developed that could transform people's lives. Scientists had created the first artificial kidney, allowing patients to live with kidney failure. Their blood was moved outside of their bodies and cleansed by a machine. Today we call it a dialysis machine, but that word didn't exist at the time. The Seattle Artificial Kidney Center had one of the first experimental machines. It was greatly oversubscribed due to the power of this lifesaving technology and its extremely limited availability.

To determine who could get this critical treatment, the head of the center, Belding H. Scribner, created a committee. It was a cross-

---

[90] He was pardoned in 2025, so maybe my startup friend was right after all.

section of society composed of seven laypeople—a lawyer, a minister, a housewife, a state government official, a banker, a labor leader, and a surgeon who served as a "doctor-citizen." The group considered the prospective patient's age, sex, marital status, net worth, income, emotional stability, nature of occupation, extent of education, past performance, and future potential to determine which of these people's lives was "worth" the most.

While Scribner's solution was a good one, it was shocking when it reached the national stage. In November 1962, *Life* magazine ran an article called "They Decide Who Lives, Who Dies."[91] While the article started as a study into this wonderful new life-saving technique, it quickly became a study of what the author referred to as The Life and Death Committee. People understood that this new technology was so powerful that it needed to be thought about differently. It wasn't just a technology question but an ethical one. This issue sparked the creation of the field of bioethics, the study of how technology applies to biology and medicine.

Real Leadership Isn't Fun and Exciting All the Time

When you're young—whether as a person or a company—leadership is about bold ideas, big risks, and making a mark. But over time, it shifts. It becomes less about proving yourself and more about responsibility. Real impact isn't always loud or dramatic—it's often found in quiet consistency, in the kind of leadership that doesn't just change the world but holds it together.

And yet, the irony of this kind of leadership is that it's difficult to celebrate. We build statues of great generals and innovators—people whose influence is visible and undeniable. But how do you honor someone whose impact comes from stepping back rather than stepping forward?

---

[91] Shana Alexander, "They Decide Who Lives, Who Dies," *Life*, November 9, 1962, http://books.google.com/books?id=qUoEAAAAMBAJ&lpg=PA102&as_pt=MAGAZINES&pg=PA102#v=onepage&q&f=false.

Sometimes we commemorate truly great, humble figures like Mother Teresa, but most of the time it feels off. Public recognition and humility don't sit easily together. We don't even have a proper word for these kinds of leaders.

Take Yale University. It's not exactly a place that screams humility. Walking through campus, past the grand Gothic buildings and towering spires, I couldn't help but feel like I'd wandered onto the set of a Harry Potter movie. The Residential Colleges—the heart of student life—were modeled after Oxford and Cambridge, and for years, I assumed they were the brainchild of some ambitious alumnus determined to make Yale just as impressive as its British counterparts. But that wasn't the story at all.

The colleges were the vision of Edward S. Harkness, Yale class of 1897. If his name sounds familiar, it's because it's everywhere on campus— Harkness Tower, Harkness Hall—though these were named after his brothers. The Harkness family helped build Standard Oil alongside John D. Rockefeller, but Edward was different. Described in George Wilson Pierson's, *Yale: The University College 1921-1937* as a "shy fat man who was unknown to many of his classmates," his freshman year at Yale was rough—few friends, a bad rooming situation, a tutor he couldn't stand. Eventually, with a little help, he found his footing, joining the fraternity Psi Upsilon and the secret society Wolfshead. But years later, watching Yale grow from 1,000 students to 1,600, he realized something: if he had gone to this bigger, busier Yale, he might not have made it.

So he did something about it. Rather than letting Yale become cold and impersonal, he used his wealth to create smaller, more intimate communities within it. It wasn't about prestige or grandeur—it was about making Yale a place where students, no matter their background, could belong. What looks today like an architectural statement was, at its core, a deeply personal gesture from someone who knew what it was like to need a little help.

And yet, despite funding the most significant transformation in Yale's history, Harkness never sought recognition. Not one of the Residential Colleges bears his name. The university tried to award him an honorary degree—three times—but his shyness always won. Once, he made it all the way to New Haven before turning around, unable to face the ceremony. In the end, his legacy wasn't about credit or acclaim. It was about quietly reshaping Yale into a place where students like him could find their way.

Final Thoughts

Amazon's final two Leadership Principles—Strive to be Earth's Best Employer and Success and Scale Bring Broad Responsibility—mark an unmistakable shift. The original principles were sharp-edged, bold, and relentlessly focused on customers. These new ones feel different. They reflect a company coming to terms with its size and influence, recognizing that leadership isn't just about moving fast and delivering results. It's also about how you treat people along the way—and what kind of legacy you leave behind.

This shift isn't just about Amazon. It's what happens to any company, or any leader, who manages to survive the early storms. In the beginning, leadership is about bold ideas and outsized ambition. Over time, it becomes something quieter and heavier. It's less about proving yourself and more about carrying the weight of responsibility. Winning isn't the finish line. It's what you do afterward that defines real leadership.

I've felt this shift in my own career. When I was younger, leadership felt personal. It was about being bold, standing out, proving something. But as I've grown, I've realized that leadership isn't about me at all. It's about what others need. That's not always easy. I've had to let go of parts of myself that once felt essential—my quirks, my instinct to challenge everything, my need to always say exactly what I think. A friend of mine, a captain in the Army, once put it bluntly: "Being a leader is about leading. You don't get to do any of the fun things anymore. That's for other people."

As a senior leader, I have a responsibility to uphold my organization's values. That doesn't mean abandoning who I am, but it does mean leadership isn't just personal expression—it's representation. People often talk about authenticity in leadership, but what does that really mean? Do you really want a CEO standing in front of a thousand people saying, "I'm in a cranky mood today because I was out too late with customers and didn't sleep well"? No. Because their mood isn't the point. Leadership at that level isn't about personal feelings—it's about setting the tone for everyone else.

I've also had to shift my focus from myself to others. When I was younger, I saw inefficiencies in organizations as absurd, almost humorously broken. I remember one time when I was working on a project that was horribly broken. It was a mess, and I kept thinking, "How is this happening? Don't they see the issues?" The problem was, I wasn't just thinking it—I was saying it. A lot. And in doing so, I was basically going around insulting people.

Everyone already knew the system was struggling. My negativity wasn't adding anything—it was just making people defensive. Eventually, I realized that if I rephrased my frustration as, "There's an enormous opportunity for improvement," I was still making the same point, but in a way that actually moved things forward—and being far less of a jerk. It shifted the focus from blame to solutions, from venting to progress. Leadership isn't about pointing out what's broken—it's about getting people to fix it.

Conclusion: My Lessons from Amazon

I learned a lot over the course of working at Amazon—from **LP #1: Customer Obsession** all the way through **LP #16: Success and Scale Bring Broad Responsibility**. But when I look back at my time there, a few lessons stand out. They've continued to influence how I see the world long after I left.

It really comes down to three things:

1. It's not about me—it's about the customer.
2. I'll never be perfect—and I don't need to be.
3. There are no shortcuts—but there are better paths.

Those ideas shaped the way I worked. But more than that, they changed how I think. What I took from Amazon goes beyond the company—unlearning old habits, trusting the process, and doing the kind of work that doesn't always feel easy, but usually ends up being right.

It's Not About Me, It's About the Customer

Before joining Amazon, I thought I understood what it meant to serve customers. I'd spent years in client-facing roles—meeting with customers, showcasing the impressive things we'd built, and explaining how those things could improve their lives. That was customer service, wasn't it?

Then I got to Amazon—and realized I had it backwards.

LP #1: Customer Obsession isn't about building something great and then convincing the customer to want it. It's about understanding what the customer actually needs—even if they can't quite articulate it—and building that. The focus shifts from "here's what I made" to "here's the problem I'm solving for you."

Earlier in my career, I would dream up ambitious, innovative solutions. I would impress my boss, and he would impress his boss, and so on. But they didn't always help the customer. One example: we built an app to help CFOs manage their corporate bank accounts. It was sleek, it was sophisticated—and it completely missed the mark. Because CFOs, as it turns out, don't actually manage operating cash. We hadn't validated the problem. We'd built something for ourselves, not for them.

At Amazon, I learned a new way of doing things.

Instead of starting with a roadmap or a list of features, you start with a press release: a clear, concise articulation of what the customer will experience once the product is built. This is the heart of Amazon's PRFAQ (Press Release and Frequently Asked Questions) process. It's a forcing function that makes you answer the hard questions before you write a single line of code. Think of it like working with a magic genie: you get one wish, and the genie will build whatever you want. It's worth taking the time to understand the vision of what you actually want him to build.

The PRFAQ is very Amazonian. It fits a culture that prizes clarity, writing, and upfront alignment. But it doesn't always travel well. When companies try to adopt the format without embracing the mindset, they end up with Frankenstein documents—PowerPoint slides pasted into Word files, six pages of jargon no one wants to read. That's not customer obsession.

Not every company is ready to write a mock press release before building a prototype. And that's okay. Because the real value of the PRFAQ isn't in the document—it's in the discipline it forces.

What matters is answering the questions:

- Who is the customer?
- What problem or opportunity are you solving for them?
- What's the most important benefit?
- How do you know what they actually need?
- What will the experience feel like when you get it right?

If you can answer those questions, it doesn't matter whether you write a PRFAQ or sketch on a whiteboard. Just don't skip the thinking. When you start with the customer, you don't just build better products—you build the right ones.

I'll Never Be Perfect, and I Don't Need to Be

Before Amazon, I saw failure as something to be ashamed of. If something didn't work, I felt like I had failed—and if it looked like it might not work, I did everything I could to make it look like it had. Success wasn't just the goal; it was the expectation.

So when I first read Amazon's **LP #4: Are Right, A Lot**—I braced myself for more of the same. I assumed it meant perfection. No room for failure. Get it right, or don't do it.

But that's not how it works.

Yes, there are decisions you can get right a lot of the time. These are the data-backed ones—the "look it up and verify it" kind. If you're choosing between two solutions and one runs twice as fast, you go with the faster one. Those are the easy calls.

But the big decisions? The ones that actually shape your business? Those are the ones where the data doesn't exist yet. You can't find the answer in the back of the book—because no one's written that chapter yet.

And that's where **LP #9: Bias for Action** comes in.

Amy Edmondson captures this beautifully with the Electric Maze exercise. The teams that win aren't the ones avoiding mistakes—they're the ones moving fast, treating every wrong step as data. They don't aim to be right; they aim to learn.

You can think of **LP #9: Bias for Action** as a kind of brute-force algorithm: try fast, fail fast, learn fast.

There's a strong temptation to hold off. To run more analysis. Schedule more meetings. Wait until you feel "ready." But the truth is, you never really feel ready. That mythical moment where everything becomes clear? It doesn't come. And the longer you wait, the harder it becomes to move forward.

It's like thanking someone for a gift. If you do it right away, a quick text is perfect. Wait a few days, and you feel like you need to write a thoughtful note. Wait a few weeks, and now it has to be a letter—or a phone call—or something really meaningful. The longer you wait, the higher the bar gets. And sometimes, you never end up doing it at all.

The same goes for action. The longer you stall, the heavier the next step becomes.

Jeff often talked about one-way and two-way doors. A one-way door is a decision that can't be undone—or at least, not easily. You want to be careful with those. But most decisions aren't like that. They're two-way doors. You can try something, see what happens, and adjust. The key is knowing the difference—and not treating every decision like you're betting your career.

For me, the shift came when I started treating more decisions like experiments—not tests. The goal isn't to prove I was right—it's to learn what's true.

I still want to be right a lot. But I've learned that the only way to get there is by moving forward, even when I don't feel ready. Not recklessly—but deliberately, with the humility to adjust.

Because I'm never going to get to the best answer by thinking. And I'm never going to get to a perfect answer at all. What I can do is act, learn, and get a little closer each time.

There Are No Shortcuts, but There Are Better Paths

I once met a group of students who had just started at an Ivy League college. They were frustrated. "I spent my whole life trying to get in," one said. "I was told that once I got here, I'd be set. But now that I'm here, it's harder than ever."

That made me think of Amazon.

People assume that once you join a company like Amazon—where the people are smarter, the execution is sharper, and the tools are more advanced—things should finally get easier. But they don't. They get more intense. That's the core of **LP #13: Have Backbone; Disagree and Commit**.

Some things do fall away. There are no endless meetings just to fill time. No busywork. No status reports designed to show that everything is on track. Thanks to **LP #10: Frugality**, there's always too much to do and never enough time to do it. You're no longer rewarded for activity. You're expected to drive outcomes. And the work that remains demands focus, judgment, and debate.

It's like taking all the disagreement that would normally show up later in the project—and pulling it forward. You talk through it early, in the open, instead of letting it simmer and drag the work down from the inside.

Sometimes your idea holds up. Sometimes someone else's does. Either way, the decision is stronger for having been discussed.

And once the decision is made, you commit. You don't hang back. You don't keep score about whose idea it was. You commit—fully—and help the team make it work.

It might not work—and that's okay. What matters is that you gave it your best shot. That's how the team learns the most: by going all in, seeing what happens, and carrying that knowledge forward. Even if it fails, you don't have to learn the same lesson twice.

This isn't easy—and it's not supposed to be. As the principle says: "They do not compromise for the sake of social cohesion." Amazon isn't trying to be a warm, fuzzy family where everyone gets along all the time. It's more like a high-performing sports team—built on trust, honesty, and shared ambition.

And that's how you get to the best answer: not by choosing the easy answer, but by refusing the easy way forward—and doing the hard work instead.

Final Thoughts

I don't follow all of Amazon's principles to the letter anymore. But some of them—especially the ones in this chapter—still shape how I think. They help me approach problems more honestly. With more curiosity. Less certainty. And a greater willingness to try, fail, and adjust.

It's the kind of mindset that stays with you—even long after you've left Seattle.

Coda: A Nice Place to Visit, but Not to Live

I spent a year at Amazon. In hindsight, it's one of the best places I've ever worked—but it's a hard place to show up to each day.

Amazon lives up to its promise. It's a company that solves hard problems in a way few others can. It's customer-obsessed, fiercely execution-driven, and relentlessly focused on improvement. I wasn't disappointed—Amazon worked exactly as advertised.

But many of the things that make Amazon a great company don't make it a great place to work. It's a highly transactional culture, where every conversation, every meeting, every decision is there to serve Amazon. At the time, there were 14 Leadership Principles—and none of them mentioned collaboration or caring about others. Instead, there's this: "Leaders do not compromise for the sake of social cohesion." That's not to say there weren't moments of camaraderie—there were. But it felt more like being on a rocket launch team: everyone sharp, aligned, and focused on the mission.

There's a lot about that culture that I admire. But over time, I realized Amazon was a great place to visit—and not a place I wanted to live. The company would help you—as long as you were helping it.

This mindset runs deep. It's baked into the company's worldview, right down to how it thinks about its own technology. In 2019, Andy Jassy—then the head of AWS—was asked about Amazon Rekognition, a facial recognition service. When questioned about U.S. customs officials using it to track undocumented immigrants, Jassy responded, "You could use a knife in a surreptitious way. There are things that you could do that you have to trust people to act responsibly with."

At Amazon, each team is a tool that's built to solve a problem for a specific customer. As emphasized in **LP #1: Customer Obsession**, success starts with understanding the customer and relentlessly addressing their needs. This principle drives Amazon's structure, breaking the organization into teams that focus on their unique challenges.

At first glance, Amazon may seem like a single unified company. But when you look closer, you'll see that each team operates independently, prioritizing its own customers. This can feel disjointed at times, especially for companies seeking bundled deals such as combining AWS services with preferred placement on Amazon.com. That's not how Amazon works. Each team is laser-focused on serving its specific customers.

This is what makes Amazon the ultimate frenemy. How Amazon operates depends largely on whether you are its customer—or its competitor. Take Tide detergent as an example. Different parts of Amazon's business interact with Tide in distinct ways:

- **As a primary distribution channel (friend):** Amazon.com sells the product through its own website. Amazon is a great retailer that ships the product to customers quickly and easily.
- **As a third-party reseller (friend):** Amazon is also letting other stores sell Tide on its website.
- **As a private-label manufacturer (enemy):** Amazon is a fierce competitor who is undercutting your prices and stealing your business by selling a similar product.

- **Overall (frenemy):** The various parts of the company are all focused on their own goals. Amazon (as a retailer) is your biggest distributor. Amazon (as a private label manufacturer) is one of your biggest competitors. The beauty of this for Amazon is that it can make money on all sides of the transaction.

It's complicated being Amazon's frenemy. You get access to their scale, infrastructure, and customers—advantages that are hard to match on your own—but you never know when they'll start competing with you.

Amazon is always in it to win it, constantly finding new ways to get ahead. Most of the time, we call it innovation. Sometimes, though, their methods feel so clever they seem close to cheating.

At the heart of it, Amazon's goal is simple: solve customer problems better than anyone else. They don't make the rules of the game—customers do. Amazon simply plays every angle, squeezing every possible advantage without technically stepping out of bounds.

Take Project Nessie, for example—a secret algorithm revealed during an FTC investigation. In the United States, it is illegal for companies to directly fix prices, such as agreeing that everyone will sell ice cream for $10. Amazon found a way to test this boundary without crossing it. Nessie would raise Amazon's prices in small ways and then watch what competitors did. If rivals followed suit, Amazon kept the higher price. If they didn't, Amazon dropped the price back down.

This wasn't outright price-fixing, but it had the same impact: raising prices for consumers while keeping Amazon technically within the rules. The FTC's recent lawsuit calls out Amazon for practices like Nessie, accusing the company of stifling competition and inflating profits at the expense of shoppers and sellers.[92]

---

[92] Federal Trade Commission, *Complaint for Injunctive and Other Equitable Relief,* September 26, 2023, https://www.ftc.gov/system/files/ftc_gov/pdf/1910134amazonecommercecomplaintrev-isedredactions.pdf.

Amazon as My Personal Frenemy

Amazon is my personal frenemy as well. It's great for me because it's very customer-obsessed—and I am the customer. It offers low prices and free, fast shipping. It gets the job done. But there is a bit more to it. Amazon's job is to help me spend my money well on its site. While it is great at getting me what I want, it cares far less about what I really need.

In the early 2000s, Jeff wrote down Amazon's core business strategy on the back of a napkin. This strategy would allow the company to dominate not just books but all of e-commerce. He called it Amazon's flywheel. It is a self-reinforcing cycle where growth in one area fuels growth in others. It's a bit like a snowball rolling down a hill—each turn adds more snow and momentum, making it bigger and faster.

For Amazon, the flywheel works by lowering prices, which attracts more customers. More customers lead to higher sales volumes, which, in turn, attract more third-party sellers to the platform, expanding product selection. The increased selection and sales volume allow Amazon to negotiate better deals and improve logistics, further lowering prices and speeding up delivery. The cycle continues, gaining momentum.

This is great for Amazon. The company takes advantage of customers' continual desire for more. Jeff says, "Customers are always beautifully, wonderfully dissatisfied, even when they report being happy and business is great." This means the company always has more to do.

But what does this mean for me? It's awful. This constant state of dissatisfaction and desire for more isn't fun. In our quest for happiness, we're hardwired to continually seek more—more wealth, greater success, and new experiences. This pursuit of more is often called the hedonic treadmill. It's baked into our DNA.

There are two things going on at the same time. First, just as our bodies maintain a set temperature, we possess a baseline level of happiness. Regardless of how many things we buy or how much we achieve, we always end up back at that baseline. So after we achieve that next new thing, our happiness gets reset. Second, we want to be happier than we are right now.

This is the relentless treadmill that causes us to constantly seek new things. Companies continue to provide us with these new things to fulfill this desire for novelty. That's how we end up with Oreos in flavors like Firework, Neapolitan, and Mocha Caramel Latte.[93]

We are so steeped in this idea of "novelty means better" that we don't realize that our novelty obsession is a relatively recent invention. In ancient times through the Middle Ages, we were focused on the past. Village elders were revered not just because they knew more than us, but because they were closer to the creation of the world, and therefore closer to God.

After the Enlightenment, things changed. As God became less central in people's lives, focus shifted to the secular world. Progress became the new ideal. Technology was creating enormous benefits and driving society forward. People were getting benefits from everything from vaccines to radio.

Then, in the twentieth century, society witnessed the horrors technology could unleash, most tragically during the Holocaust. People became disillusioned with the idea that technology would make the world a better place. Instead of progress, society fell back on the idea of innovation—novelty for its own sake.

You might think, *Why can't I have both? Why can't I get more things now and have them make me happy in the future?* The hedonic treadmill is one answer. But also consider that the pursuit of happiness in the future makes you less happy today. As humans, we have an inborn equation for satisfaction:

Satisfaction = What You Have ÷ What You Want

So while Amazon is obsessed with you, their customer, they are obsessed in a very specific way. They want to help you run faster on that hedonic treadmill.

---

[93] I love picking on Oreos' innovation. Googling "Oreo innovation" brings up great articles like "The Business Strategy Behind Oreo's Constant, Weird New Flavors."

Taking Control

Amazon is amazing at giving us exactly what we want, right when we want it. But its real goal goes deeper—it's trying to rewire our brains. That might sound a little sinister, but it's really what any successful company is after: becoming your default solution. You think, "I need something," and Amazon wants your next thought to be, "I'll just get it right now from Amazon." It's a brilliant strategy that keeps us hooked. But what's great for Amazon isn't always great for us.

Amazon's secret lies in how it taps into our impulsive side. One-click purchasing, personalized recommendations, and next-day delivery make it effortless to act on a whim. It's a system designed to convince us that buying more stuff will make us happy. And it works—we keep reaching for that next thing, hoping it will fill the gap. But most of the time, what we really need—connection, purpose, or peace of mind—can't be delivered in a cardboard box.

Our brains have two sides. One is fast, impulsive, and reactive—what I like to call the "Amazon side." The other is slower and more deliberate. That's the side that pauses and asks, *Do I really need this?* It's where we think about trade-offs, long-term goals, and what actually makes life better. But Amazon's entire system is built to keep us from tapping into that reflective part of ourselves. It keeps us locked in the here-and-now, chasing instant gratification that feels good in the moment but often leaves us empty later.

And it's not just Amazon. Society reinforces this mindset, constantly telling us we need more—more stuff, more prestige, more success. Resisting this pressure is hard. It feels like going against everything we've been taught.

Stepping off the treadmill is hard. The world around us is constantly pushing us to keep running. But even in the most relentless circumstances, Viktor Frankl showed that people still have the power to choose how they respond.

Frankl, a Holocaust survivor and author of *Man's Search for Meaning*, lived through unimaginable suffering. Stripped of his freedom, dignity, and even his family, he uncovered a profound truth: no matter how much is taken from us, one freedom remains. "Everything can be taken from a man but one thing," he wrote, "the last of the human freedoms—to choose one's attitude in any given set of circumstances, to choose one's own way."

Frankl witnessed this power of choice in the darkest places. Even in concentration camps, where prisoners were dehumanized and starved, some chose acts of kindness—offering comfort or sharing the smallest scraps of food. These moments, though small, were profound. They proved that even in the most harrowing circumstances, we can still decide how we respond.

If Frankl could make that choice in the face of such horrors, surely we can make it today. Society pressures us to conform, consume, and endlessly chase after more, but we can pause and ask, "What truly matters to me?" As Paul Graham noted, "Prestige is what other people want you to do," and if we're operating mostly on impulse, we might find ourselves chasing achievements that don't truly resonate with our values.

By taking control back from the world—from Amazon, the tech companies, and society's relentless demands—we can begin to use these forces as tools, rather than letting them use us. The choice is ours, and in that choice lies the power to shape a meaningful life.

An Ideal Retirement

Let's think about retirement—not the glossy fantasy, but what it really means. If you start from the end and work backwards, you can ask yourself, *What do I want my life to actually look like?* For years, I thought I knew the answer.

I knew that money wouldn't make me happy, but I still envied the people who could retire early. I dreamt of being that person who quit their job, moved to Hawaii, and sipped margaritas while I cashed my dividend checks. But as I got older, I realized that it's not about the age of retirement but the quality of it.

But then, in 2004, I met a 30-year-old who had just sold his company to Amazon. He was a nice guy, so my wife and I were trying to set him up with some of our friends. How could this go wrong? Who wouldn't want a nice Jewish rich man who retired at 30? The only problem was that he acted like a retiree who'd just moved to Florida. He didn't go out much, watched a lot of TV, and volunteered at some very noble charities delivering food. Though he was rich, no one wanted to date him because he lived a very boring life.

There are early retirees whom I do envy. The people who have enough money to choose to do what's important. Whether it's my friend who retired from Facebook at 40 to mentor other women or the dad who sold his company and can now volunteer at preschool. They will work, even if it's not the standard definition of a 40+ hour workweek.

But it's my friend Heath who really blew my mind. One day he got a cold call from a money manager.

The money manager said, "Hi, Heath, I'm Bob, and I can make your life much better. I'm a licensed financial planner and can help you retire much sooner. My clients are able to retire five years earlier with my planning."

Heath said, "I think you have the wrong person. I don't need your services."

The money manager said, "Do you have someone else? Because I have great rates and a phenomenal track record."

"No," Heath said, "you don't understand me. More money won't help me. I love my job. I'm a doctor and I work four days a week. This gives me time to spend with my three kids and also to deliver real value for my patients. I have enough money to support my lifestyle and I love what I do."

He said, "But wouldn't it be better to retire early and work less?"

Heath said, "Again, you're missing my point. I can't work less than four days a week and serve my patients well. I have no interest in retiring. I enjoy helping my patients lead better lives, and the world needs more doctors seeing patients, not playing golf."

That's why I envy Heath the most. He is not working to retire. He is working because he has already found the life he wants.

Retirement is the ultimate freedom, but it's really just a means to an end. Freedom—and the money that gets you there—isn't the point. As the tech publisher Tim O'Reilly once said, "Money is like gas in the car—you need to pay attention or you'll end up on the side of the road—but a successful business or a well-lived life is not a tour of gas stations."

Final Thoughts

We've talked about the Leadership Principles and Working Backwards. We've delved into Two Pizza Teams and Disagree and Commit. I've even shared how I've taken these ideas and used them in my own life. With all this knowledge, it might seem like we're ready to take over the world.

But that's not the real lesson from this journey. The truth is, there are no easy answers. Every choice comes with trade-offs. Throughout my career, I've worked with many companies that dream of shortcuts to becoming the next Amazon. They want to hire a consultant, reorganize their teams, and—like magic—transform overnight. They want the innovation of an Amazon Go store, but without any of the failures that come with it.

I can't buy a pre-packaged solution to all of my problems. I can buy better and better tools—sharper, more powerful, and more personalized tools. These tools can give me superpowers to do more, but they don't necessarily make life better. It's all about how I use them.

Remember, don't take all of this too seriously. The goal of this book is to help you achieve your goals and find more fulfillment. Try these ideas on for size and see what works, but don't overdo it. While these are best practices, they won't fit every case or every person—they're just averages.

Nassim Nicholas Taleb, the contrarian trader from **LP #8: Think Big**, is skeptical of averages. In his book *Skin in the Game*, he introduces the concept of ergodicity, explaining that averages are just that—they don't apply to every situation. Take Russian roulette, for instance. On average, you win five out of six times. But that does not make it a good idea.

So don't take this stuff too seriously. Listen to Louis Menand, the New Yorker writer. Menand was writing about the book *The Power of Habit* by Charles Duhigg. In it, Duhigg took on a personal project of self-improvement. He noticed that he would eat a cookie in the middle of the afternoon. Wanting to change this behavior, Duhigg analyzed what prompted this cookie break and replaced it with something more productive—a chat with his colleagues. This prompted Menand to say the following:

> He soon found that he no longer needed the cookie. He had management-theorized himself into becoming a more disciplined person. The story made me sad. Mr. Duhigg. Charles. Life is short. Eat the cookie.[94]

---

[94] Louis Menand, "The Life Biz," *The New Yorker*, March 28, 2016, https://www.newyorker.com/magazine/2016/03/28/smarter-faster-better-the-secrets-of-being-productive-in-life-and-business.

APPENDIX A
THE LEADERSHIP PRINCIPLES[95]

We use our Leadership Principles every day, whether we're discussing ideas for new projects or deciding on the best way to solve a problem. It's just one of the things that makes Amazon peculiar.

Customer Obsession

Leaders start with the customer and work backwards. They work vigorously to earn and keep customer trust. Although leaders pay attention to competitors, they obsess over customers.

Ownership

Leaders are owners. They think long term and don't sacrifice long-term value for short-term results. They act on behalf of the entire company, beyond just their own team. They never say "that's not my job."

Invent and Simplify

Leaders expect and require innovation and invention from their teams and always find ways to simplify. They are externally aware, look for new ideas from everywhere, and are not limited by "not invented here." As we do new things, we accept that we may be misunderstood for long periods of time.

Are Right, A Lot

Leaders are right a lot. They have strong judgment and good instincts. They seek diverse perspectives and work to disconfirm their beliefs.

Learn and Be Curious

Leaders are never done learning and always seek to improve themselves. They are curious about new possibilities and act to explore them.

---

[95] Amazon, "Leadership Principles," https://www.amazon.jobs/content/en/our-workplace/leadership-principles.

Hire and Develop the Best

Leaders raise the performance bar with every hire and promotion. They recognize exceptional talent, and willingly move them throughout the organization. Leaders develop leaders and take seriously their role in coaching others. We work on behalf of our people to invent mechanisms for development like Career Choice.

Insist on the Highest Standards

Leaders have relentlessly high standards — many people may think these standards are unreasonably high. Leaders are continually raising the bar and drive their teams to deliver high quality products, services, and processes. Leaders ensure that defects do not get sent down the line and that problems are fixed so they stay fixed.

Think Big

Thinking small is a self-fulfilling prophecy. Leaders create and communicate a bold direction that inspires results. They think differently and look around corners for ways to serve customers.

Bias for Action

Speed matters in business. Many decisions and actions are reversible and do not need extensive study. We value calculated risk taking.

Frugality

Accomplish more with less. Constraints breed resourcefulness, self-sufficiency, and invention. There are no extra points for growing headcount, budget size, or fixed expense.

Earn Trust

Leaders listen attentively, speak candidly, and treat others respectfully. They are vocally self-critical, even when doing so is awkward or embarrassing. Leaders do not believe their or their team's body odor smells of perfume. They benchmark themselves and their teams against the best.

Dive Deep
Leaders operate at all levels, stay connected to the details, audit frequently, and are skeptical when metrics and anecdote differ. No task is beneath them.

Have Backbone; Disagree and Commit
Leaders are obligated to respectfully challenge decisions when they disagree, even when doing so is uncomfortable or exhausting. Leaders have conviction and are tenacious. They do not compromise for the sake of social cohesion. Once a decision is determined, they commit wholly.

Deliver Results
Leaders focus on the key inputs for their business and deliver them with the right quality and in a timely fashion. Despite setbacks, they rise to the occasion and never settle.

Strive to be Earth's Best Employer
Leaders work every day to create a safer, more productive, higher performing, more diverse, and more just work environment. They lead with empathy, have fun at work, and make it easy for others to have fun. Leaders ask themselves: Are my fellow employees growing? Are they empowered? Are they ready for what's next? Leaders have a vision for and commitment to their employees' personal success, whether that be at Amazon or elsewhere.

Success and Scale Bring Broad Responsibility
We started in a garage, but we're not there anymore. We are big, we impact the world, and we are far from perfect. We must be humble and thoughtful about even the secondary effects of our actions. Our local communities, planet, and future generations need us to be better every day. We must begin each day with a determination to make better, do better, and be better for our customers, our employees, our partners, and the world at large. And we must end every day knowing we can do even more tomorrow. Leaders create more than they consume and always leave things better than how they found them.

APPENDIX B
INTERVIEWING AT AMAZON

If you're reading this book because you want to work at Amazon, you're in the right place—at least to start. This book will give you a solid high-level understanding of the company and its culture. But I wouldn't rely on it as your only (or even your main) resource to prepare for the interview.

A lot of what's in these pages is personal: how I experienced the Leadership Principles, how I applied them outside Amazon, and what I took with me afterward. That context is valuable—but your interview will be focused squarely on work. Think meat-and-potatoes examples, not personal reflections.

While I give a general overview of the interview process in **LP #6: Hire and Develop the Best**, this section will go deeper into how to prepare your stories, understand what interviewers are listening for, and build the kind of answers that actually land.

One quick tip: internal referrals matter. At Amazon, I got a steady stream of requests for informational interviews—usually from people hoping I'd refer them. Eventually, I realized it was much easier (and honestly more helpful) to just give the referral up front rather than spend 30 minutes on a conversation that was really just a pretext. So I started saying, "I'm happy to refer you. Once you're selected for the interview, we can do the informational interview."

It might sound a bit blunt, but it's not rude—it's just efficient. Once someone landed the interview, that's when I could actually be helpful: reviewing their stories, helping them prepare, and offering targeted advice—like what you'll find in the sections below.

What to Read?

If you're preparing for an Amazon interview, start with the "Further Reading" section at the end of this book. Don't treat it like a list to memorize—it's there to help you understand how Amazon thinks.

Start with the Leadership Principles. You need to know these cold. The interviewers will use these to measure how Amazonian you are. But don't just memorize the list. Notice how the principles overlap, how they build on each other—and sometimes even contradict each other. That's by design. It's your job to know when to use the appropriate principle and how to apply it.

Then spend some time on Amazon's "How We Hire" page. It lays out what the process will look like.

For a deeper dive, check out *Scarlet Ink*, a blog by Dave Anderson, a former Amazon Bar Raiser. His posts break down each Leadership Principle and explain what interviewers are actually listening for. It's practical, sharp, and unusually honest. I found it invaluable when I was preparing.

Working Backwards is the best general book on how Amazon operates—not for interview tips, but for context. It explains how tools like the PRFAQ and six-pager came to be, and why the culture feels so peculiar to outsiders. If you want to sound like someone who understands Amazon's internal logic, start here.

The rest—the shareholder letters (especially the 1997 one and the most recent one), the AWS innovation whitepaper, biographies, and internal videos—won't directly help you pass the interview. But they'll give you a feel for the mindset. And ultimately, that mindset is what Amazon is hiring for.

Preparing Your Stories

You'll need a set of stories for your interview.

Amazon interviewers are trained to evaluate your answers using the STAR method. It's the structure they're listening for—and the format they'll use to take notes. STAR is simple, but powerful:

- **Situation** – set the scene
- **Task** – explain what needed to be done

- **Action** – describe what you did
- **Result** – share what happened

Think of it as telling a short, focused story with a clear arc. STAR helps you go beyond surface-level claims like "I'm great at ownership" and actually show it—by walking through a moment where you owned a problem and delivered a measurable result. It forces you to be specific, and that's what makes your story stick.

Keep your answers crisp. If you're going past two minutes, you're probably wandering.

Each interviewer is assigned two or three Leadership Principles. As you speak, they'll be listening for evidence that you embody those specific LPs—and capturing your story in STAR format. That's why it's important to come in with a few core stories that you can adapt to different questions. Here's how I coach people to develop their core list of stories:

1. Start by understanding the Leadership Principles and Amazon's culture, as discussed earlier in this chapter. That foundation matters—it helps you choose the right stories and frame them in an Amazonian way.

2. For each Leadership Principle, brainstorm one or two stories that show how you've demonstrated that principle. Focus on moments you're genuinely proud of—times when you took ownership, delivered results, or solved a tough problem. These should be specific, personal wins.

3. Create a table to organize your thinking. List the Leadership Principles down the left-hand side and your potential stories across the top.

4. Score each story from 1 (low) to 5 (high) for how well it represents each principle. A story can appear in more than one row if it hits multiple principles.

5. Watch out for stories that are "Un-Amazonian." Some stories may align with one principle but violate others. For example, you might have gone deep on a technical rabbit hole and learned a lot, but missed a major deadline and delivered no customer value. If you can't find a way to fix those problems, they should be scored as a negative example—call it a -∞.
6. Add up the scores for each story.
7. The highest-scoring stories become your core Amazon stories. These are the ones that can flex across multiple principles and questions. You don't want to repeat the same story with the same interviewer, but it's perfectly fine to reuse strong stories across different interviewers—just emphasize different angles depending on the question.

LP	Story 1	Story 2	Story 3	...
LP 1	4	1	5	
LP 2	5		5	
LP 3	3	5	-∞	
...				
Total	12	6	Don't Use	

Example Story Scoring Grid

It's important to have solid Amazonian stories that clearly reflect the Leadership Principles. You don't need one story for every LP—but you do need five or six strong, versatile stories that can cover multiple principles. Here's why:

1. You want to avoid telling a story that accidentally goes against another principle.
2. You won't know in advance which principles your interviewer is assigned.
3. Once you figure out which LPs they're focused on, you'll want to pivot to a story that matches.
4. A few strong, well-rehearsed stories are more effective than a long list of one-off examples.

For example, I have a great story about mentoring—but it really only covers mentorship. I also have a slightly less polished story that still touches on mentorship, but also demonstrates **LP #1: Customer Obsession** and **LP #8: Think Big**. I usually go with that one. My strongest story hits eight of the LPs.

Stress Test Your Stories

Many companies are secretive about their interview questions. Amazon isn't. The questions are fairly standard and easy to find with a quick search for "Amazon Leadership Principles interview questions." If you're at a school, your career center probably has a list as well.

Go through these questions and see how your stories line up. A strong Amazon story should be flexible enough to answer multiple questions with a bit of tailoring. It might take some effort to map each story to a specific question, but if the story is solid, it will fit.

In the actual interview, aim to tell the best-matching story you have—even if it's not a perfect fit. A well-practiced story that hits key principles will almost always land better than something you try to make up on the spot.

Build your confidence by practicing different versions of questions tied to the same principle. They may sound different, but they're getting at the same core behavior. For example, Interview Genie offers several variations of questions for the **LP #12: Dive Deep** principle:

- Have you ever leveraged data to develop strategy?
- Tell me about a time you were trying to understand a problem on your team and you had to go down several layers to figure it out. Who did you talk with and what info proved most valuable? How did you use that info to help solve the problem?
- Tell me about a problem you had to solve that required in-depth thought and analysis. How did you know you were focusing on the right things?

Make Sure You Add Enough Detail

Make sure you can answer each question with enough detail. Just because you know everything you did doesn't mean your interviewer will. You have to spell it out clearly.

One helpful tool is the "Five Whys" approach we discussed in **LP #12: Dive Deep.** Keep asking "why" until you get down to the core insight or decision-making process that shaped your actions.

Here's an example from a conversation I had with my mentee Sam while she was preparing for her Amazon interview. By pushing herself to go deeper, Sam was able to clarify and strengthen her story. She ended up using it in the interview—and she got the job.

> **Me:** Tell me about a time when you had to Dive Deep to understand an issue.
>
> **Sam:** I was a consultant on a big merger of two healthcare companies, an insurance company and a pharmacy benefits manager. The two companies were having issues because they had very different cultures, mainly because one was much bigger and the companies had different business models. I conducted interviews and focus groups with different groups of people, from the C-Suite to the front-line workers at the drug manufacturing sites. I used data analysis to identify culture gaps and mismatches and come

up with a solution to address the behavioral changes to support the merger. Eventually we were able to come up with common ways of working that satisfied both companies.

At this point, Sam felt confident in her answer. She'd described multiple ways she approached the problem. But she was still skimming the surface.

Me: Companies always have different cultures. What did this look like at these particular companies?

Sam: Let's call them Big Co. and Small Co. Big Co. was long-established, highly process-driven, and very risk-averse. Small Co. moved quickly and embraced a "fail fast" mindset. The mismatch was creating tension—especially for Small Co. employees who felt slowed down.

OK, now we're starting to get somewhere.

Me: How did you know this?

Sam: I would see the frustration from Small Co. employees during focus groups.

Me: Tell me an example.

Sam: Sure. Mark, an account executive, told me he used to have complete autonomy in pricing negotiations. Now, after the merger, he needed approval for any discount over 10%—and he had to get three or four of those a day. It was taking up hours of his time and hurting his ability to build client relationships.

Me: So what did you do to help him?

Sam: I realized the issue wasn't just with Mark. Even Big Co. employees were frustrated with all the approvals and meetings. I proposed a new approval workflow that clarified who needed to be involved, streamlined meeting participation, and delegated more pricing authority to frontline staff—while still meeting Big Co.'s risk requirements. The company implemented the new system, and it helped reduce friction across the board.

Sam's story worked because she didn't stay at the surface. We kept peeling back the layers until she uncovered the real problem— and showed how she solved it in a way that balanced competing priorities. Each follow-up revealed more depth: the cultural tension, the operational bottlenecks, and the human impact behind them. By the end, she wasn't just describing a project—she was showing how she thought, how she navigated complexity, and how she delivered results. That's what Amazon interviewers are listening for.

At the end of the day, interviewing at Amazon is about preparation, clarity, and depth. You don't need to be perfect—but you do need to be thoughtful, specific, and grounded in real work you've done. If you've made it this far in the book, you're already thinking like an Amazonian. Keep going. Practice your stories, ask yourself the hard questions, and trust your experience.

Good luck in your interviews!

APPENDIX C
A FEW OF MY FAVORITE PECULIAR THINGS

Here are a few of some of my favorite things I learned at Amazon:

- **Funny Product Reviews:** At the end of my onboarding, the final assignment was to read some humorous product reviews that showcase Amazon's quirky, peculiar culture.[96] While Amazon is meticulous about removing misleading reviews, it embraces the intentionally humorous ones. The company even produced an audio version of these reviews read by celebrities, which feels a bit clunky.[97] You've got to give them credit for trying, though! Some of my favorites include:
  - BIC Cristal For Her Ball Pen: "Is it safe for my husband to use?"
  - Hutzler 571 Banana Slicer: "Get this if you've ever been frustrated by the chore of slicing bananas!"
- **Amazon's Building Names:** Amazon names its buildings using the codes of nearby airports, a nod to its e-commerce roots and vast fulfillment network. I worked in JFK14. Many packages in New York come through JFK8, the Staten Island Fulfillment Center. Some buildings, especially those in Seattle, also have a nickname. Jeff works in Day 1. Other names include:
  - Rufus (named after the first company dog)
  - Doppler (the code name for Amazon Echo)
  - Nessie (named after the Loch Ness Monster)
- **The Amazon Mothership:** Unlike many tech companies that opt for suburban campuses, Amazon is based in downtown Seattle. Employees and visitors can take a self-guided tour of the campus.[98]

---

[96] There are a lot of articles out there about fake reviews, but one of my favorites is an early piece from 2013—Maria Popova's essay on Brain Pickings called "Modern Masterpieces of Comedic Genius: The Art of the Humorous Amazon Review." It's one of the first to treat these reviews not just as jokes, but as tiny works of literary art. https://www.themarginalian.org/2013/07/08/humorous-amazon-reviews/

[97] Jane Lynch (narrator), *Funny Amazon Reviews*, Audible audiobook, 2017, https://www.audible.com/pd/B071WXMZR3?source_code=ASSORAP0511160006.

[98] Amazon Tours Puget Sound, https://pugetsound.amazontours.com.

There's even an audio version. A few notable stops include:

- The Spheres, a giant terrarium filled with plants designed to inspire creativity. It is open to the public only on certain days and times.
- Amazon Go store to experience Amazon's "Just Walk Out" cashier-less future.

- **The Amazon Product Graveyard:** Like any inventive company, Amazon has had its fair share of flops. There's even an unofficial "cemetery" of discontinued experiments, lovingly curated by fans of big swings and bigger misses. It's nice to remember that being the best place to fail has some real-world roadkill. Some of my favorite headstones include:
  - **Treasure Truck:** This was Amazon's version of a daily-deal party on wheels. You'd get a ping on your phone—"Today's Treasure Truck deal is ribeye steaks!"—and race to a truck parked somewhere nearby to pick it up. Part flash sale, part scavenger hunt, part marketing fever dream. I'm not sure what its long-term plans were, but it was delightfully weird while it lasted.
  - **Amazon Dash Buttons:** A physical button you could stick next to your washing machine or toilet paper stash. Press it, and Amazon would reorder the item for you—no screen required. Amazon soon realized that a better way to get you more paper towels was through Subscribe & Save.
  - **Amazon Spark:** Amazon tried to build a social feed that felt like Instagram, but with shopping. Users posted photos of their favorite products, and others could buy straight from the post. It was niche, peculiar, and short-lived—turns out people don't scroll through lifestyle pics to shop for extension cords. It was a good try. Social shopping worked well for Venmo.

APPENDIX D
GLOSSARY

Amazonians: Employees who work at Amazon. The term reflects a shared identity and culture among people across the company.

Andon Cord: A mechanism that allows employees to stop a process when a problem is detected. Adapted from Toyota, it helps prevent defects from progressing and promotes immediate problem-solving.

AWS (Amazon Web Services): A subsidiary of Amazon that provides cloud computing services, allowing companies to rent servers, storage, and other infrastructure as needed.

Bikeshedding: The tendency to focus on trivial decisions while avoiding more complex or impactful ones, often because the simple tasks feel more manageable.

Cargo Cult Thinking: The act of copying visible processes without understanding their purpose or function, leading to ineffective or superficial results.

EC2 (Elastic Compute Cloud): A service within AWS that provides scalable virtual servers, allowing customers to run applications in the cloud without maintaining physical machines.

Fulfillment Center: A warehouse where Amazon stores, picks, packs, and ships items to customers.

Jeff: Jeff Bezos, the founder of Amazon. Referred to by his first name throughout the book, as is common within Amazon's internal culture.

Jeff's Shadow: The Technical Assistant to the CEO, who observes and supports Jeff in high-level meetings.

Leadership Principles: Sixteen core values that guide behavior, decision-making, and performance at Amazon. They define what it means to be an Amazonian.

LP: Abbreviation for Leadership Principle. Frequently used internally at Amazon.

Mechanism: Amazon's word for a lightweight Amazonian process.

Minimum Viable Product (MVP): The simplest version of a product that addresses a customer problem. Used to test assumptions quickly and guide further development.

Obidos: Amazon's original website codebase, named after a fast-moving section of the Amazon River. Eventually replaced due to scalability limitations.

Press Release: A fictional announcement written at the start of a project to define the customer value of a product. It guides development by clarifying goals.

PRFAQ (Press Release and FAQ): A planning document that combines a press release with a list of anticipated Frequently Asked Questions and their answers. Central to Amazon's Working Backwards process.

S3 (Simple Storage Service): An AWS service that provides scalable object storage. Used by companies to store and retrieve data on demand.

Tenet: A guiding principle that helps a team or organization make consistent decisions and prioritize what matters most.

Two-Pizza Teams: Amazon's preferred team size—small enough to be fed with two pizzas. Encourages autonomy, speed, and clearer communication.

APPENDIX E
FURTHER READING

Books About Amazon

These are some of my favorite books about Amazon. They fall into two categories: insider accounts and outside perspectives. Brad Stone's books are independent and well-researched, though Jeff famously declined to participate. (His then-wife, MacKenzie Scott, left *The Everything Store* a one-star review.)

- **The Everything Store** by Brad Stone – A journalistic look at Amazon's early days and rise to dominance.
- **Amazon Unbound** by Brad Stone – The sequel to *The Everything Store*, covering Amazon's expansion into AWS, Alexa, and beyond.
- **Working Backwards** by Colin Bryar and Bill Carr – Written by two longtime Amazon executives, this book explains Amazon's internal tools and practices from the inside.
- **The Amazon Way** by John Rossman – Written by a former Amazon executive, this book introduces the original 14 Leadership Principles and shows how they shape decision-making and innovation.

Jeff's Writing

Jeff doesn't do much press, but like any good Amazonian, he writes a lot. These selections show how he thinks about long-term strategy, innovation, and culture.

- **Shareholder Letters** – Start with the 1997 letter—it's the original and still the blueprint. From there, wander through the rest of Jeff's letters and you'll find plenty of Amazonian thinking:

Customer Obsession, Day 1, high standards, and more. Jeff's annual letters are a running commentary on how Amazon thinks, grows, and makes decisions. https://ir.aboutamazon.com/annual-reports-proxies-and-shareholder-letters/default.aspx

- **Invent and Wander by Jeff Bezos, with an introduction by Walter Isaacson** – This includes Jeff's annual letters, combined with select speeches, essays, and interviews. https://www.amazon.com/dp/B08BCCT6MW

Amazon Materials

These documents come straight from Amazon and offer a clear look into how the company actually operates.

- **Amazon's Leadership Principles** – The official Amazon page on Leadership Principles, including video explanations by Andy Jassy. https://www.amazon.jobs/en/principles
- **AWS Culture of Innovation Whitepaper** – A nice overview of Amazon's culture and the Working Backwards process. https://pages.awscloud.com/rs/112-TZM-766/images/EV_aws-culture-of-innovation_Jul-2020.pdf

Interview References

If you're preparing for an Amazon interview, here are some great resources to dig into.

- **Amazon's Hiring Page** – Includes an overview of the hiring flow, behavioral interviews, and the STAR framework (Situation, Task, Action, Result). https://www.amazon.jobs/content/en/how-we-hire
- **Scarlet Ink** by Dave Anderson – A former Bar Raiser and Amazon tech director, Dave has built a thoughtful library of

interview guides. His breakdown of the Leadership Principles was especially helpful when I was preparing to interview. https://www.scarletink.com/t/interviewing

Related Books

These aren't about Amazon specifically, but they helped shape the way I think about risk, learning, and leadership—and they show up throughout this book.

- **The Black Swan** by Nassim Nicholas Taleb – This book was on Jeff's senior management reading list. A book about how rare, unpredictable events shape the world more than we like to admit—and why most of us are terrible at preparing for them.
- **Antifragile** by Nassim Nicholas Taleb – The follow-up to *The Black Swan*. It explores how certain systems (and people) get stronger when exposed to stress, disorder, and uncertainty.
- **Mindset** by Carol S. Dweck – This is the classic book on the growth mindset, showing how we can learn to love growth instead of hating failure.
- **The Right Kind of Wrong** by Amy C. Edmondson – Similar to Dweck's book, but this one looks at organizations. Edmondson explains why most organizations fear failure—and how to build systems that learn from it instead.

Jeff's Videos

It's fun to watch Jeff on video throughout the years. These videos capture some of his clearest—and sometimes quirkiest—thinking about leadership, innovation, and Amazon's evolution.

- **A Young Jeff on 60 Minutes** (1999) – An early interview with Jeff back when he still had hair—and a wild glint in his eye.

He talks about leaving a stable job to start Amazon, taking big risks, and building something new on the internet. https://www.youtube.com/watch?v=InEgmXhU4KA

- **Jeff Bezos TED Talk: The Electricity Metaphor** (2007) – The dot-com boom and bust is often compared to the Gold Rush, but Jeff says it's more like the early days of the electric industry.

- **An Early View into AWS** (2008) – A very casual interview with Om Malik on the founding of Amazon Web Services. Jeff explains how AWS was built to solve Amazon's own scaling problems and eventually became a platform for everyone else. https://videopress.com/v/o9nEqEtT

- **Jeff Bezos at the Economic Club of Washington** (2018) – The video has many of the best Jeffisms, including Day 1, regret minimization, and why Customer Obsession is the cornerstone of Amazon's strategy. A great overview of his leadership philosophy. https://www.youtube.com/watch?v=zN1PyNwjHpc

Acknowledgments

When I was at Amazon, we gave out awards for exceptional use of the Leadership Principles—usually a superhero figurine with the principle emblazoned on it.

In that same spirit, I'm handing out a few Leadership Principle awards of my own. There's no figurine this time, but these are the people I want to recognize for helping me write this book and bring it to life.

- **Earn Trust:** My family, for trusting that behind all the pacing, scribbling, and "I'll be right there, just one more paragraph," something worthwhile was actually being built.
- **Ownership:** Sameer C., who didn't just cheer me on—he showed me the whole publishing maze and handed me a map.
- **Customer Obsession:** Seth G., for being the customer who most wanted this book to exist—even if it meant pestering the author into submission.
- **Hire and Develop the Best:** Tom Fiffer and his team at Christmas Lake Creative, who helped sharpen every sentence and encouraged me to keep the bar high (and the typos low).
- **Insist on the Highest Standards:** Pete A., Troy A., Amber J., Heski B., Peter T., and Lutz B.—early readers who didn't just read drafts, but made them better.

- **Dive Deep:** Abi R., Adrian B., Christine A., Dan S., Deepthi G., Eric B., Jeanne Z., Jen S., Jonathan H., Kavita J., Lee H., Marc E., Mike K., Nicole P., Ravi S., Rebecca E., Ros R., Shana R., Stacey C., Taira H., Tera H., Vali T., and the rest of the friends who aggressively read my stuff, pointed out the gaps, and encouraged me to keep going.
- **Learn and Be Curious:** Collaborative Gain—especially Council 9 and my Writing Roundtable—for giving me a space to keep learning, questioning, and thinking out loud.
- **Think Big:** Raja R., Rich R., Peter T., Tera H., Seth G., and Sameer C.—for putting *Peculiar* into worlds I never imagined: boardrooms, classrooms, chaos labs, hockey arenas, sales teams, and even fantasy novels.

www.ingramcontent.com/pod-product-compliance
Lightning Source LLC
Chambersburg PA
CBHW031504180326
41458CB00044B/6693/J